T0384037

Seamless 3D Navigation in Indoor and Outdoor Spaces

This book presents the current research on space-based navigation models and the contents of spaces used for seamless indoor and outdoor navigation. It elaborates on 3D spaces reconstructed automatically and how indoor, semi-indoor, semi-outdoor, and outdoor spaces can mimic the indoor environments and originate a network based on the 3D connectivity of spaces. Case studies help readers understand theories, approaches, and models, including data preparation, space classification and reconstruction, space selection, unified space-based navigation model derivation, path planning, and comparison of results.

Features:

- Provides novel models, theories, and approaches for seamless indoor and outdoor navigation path planning
- Includes real-life case studies demonstrating the most feasible approaches today
- Presents a generic space definition framework that can be used in research areas for spaces shaped by built structures
- Develops a unified 3D space-based navigation model that allows the inclusion of all types of spaces as 3D spaces and utilizes them for seamless navigation in a unified way

Intended to motivate further research and developments, this book suits students, researchers, and practitioners in the field, and serves as a helpful introductory text for readers wanting to engage in seamless indoor/outdoor navigation research and teaching.

Seamless 3D Navigation in Indoor and Outdoor Spaces

Jinjin Yan and Sisi Zlatanova

CRC Press is an imprint of the
Taylor & Francis Group, an **informa** business

First edition published 2023
by CRC Press
6000 Broken Sound Parkway NW, Suite 300, Boca Raton, FL 33487-2742

and by CRC Press
4 Park Square, Milton Park, Abingdon, Oxon, OX14 4RN

CRC Press is an imprint of Taylor & Francis Group, LLC

ISBN: 978-1-032-24664-2 (hbk)
ISBN: 978-1-032-25002-1 (pbk)
ISBN: 978-1-003-28114-6 (ebk)

DOI: 10.1201/9781003281146

Typeset in font Nimbus
by KnowledgeWorks Global Ltd.

Publisher's note: This book has been prepared from camera-ready copy provided by the authors.

Contents

Preface ... ix

About the Authors ... xi

Chapter 1 Introduction ... 1

1.1 Navigation Concepts ... 1
1.2 Current Attempts on Seamless Navigation ... 3
1.3 Notions and Terminology ... 6
1.3.1 Five Existing Concepts ... 6
1.3.2 Six New Terms for Built Structures ... 11
1.3.3 Five New Terms for Space ... 15
1.4 Book Overview ... 18

Chapter 2 Spaces for Seamless Navigation ... 21

2.1 Space Definitions ... 21
2.2 Space Classification ... 25
2.2.1 Living Environments ... 25
2.2.2 Indoor & Outdoor ... 26
2.2.3 Semi-bounded Spaces ... 27
2.2.4 Four Examples of Space Classification and Definition Framework ... 30
2.3 Space Representation ... 31
2.3.1 BReps ... 32
2.3.2 Voxels ... 32
2.3.3 Examples of Space Geometric Representations ... 33
2.4 A Generic Spaces Definition Framework ... 35
2.4.1 Descriptive Definition ... 35
2.4.2 Quantitative Definitions ... 37
2.4.3 Illustration of the Generic Space Definition Framework ... 38
2.5 Summary ... 44

Chapter 3 Space-based Navigation Models ... 45

3.1 Navigation Network ... 45
3.1.1 The Poincaré Duality Theory ... 46
3.1.2 Approaches for 2D Navigation Network Derivation ... 46

3.1.3 Approaches for 3D Navigation Network Derivation ... 48
3.2 International Standards Related to Navigation ... 49
3.2.1 IndoorGML ... 49
3.2.2 Industry Foundation Classes (IFC) ... 51
3.2.3 CityGML ... 51
3.3 Navigation Network Derivation for QR Code-based Indoor Navigation ... 53
3.3.1 QR Code-based Indoor Navigation ... 53
3.3.2 Indoor Scene Classification ... 55
3.3.3 Space Subdivision and Navigation Network Derivation ... 57
3.3.4 Dummy Nodes and Extended Navigation Network ... 58
3.4 Summary ... 59

Chapter 4 Unified Space-based Navigation Model ... 63

4.1 Requirements to a Unified Space-based Navigation Model ... 63
4.2 Conceptual Model of Unified 3D Space-based Navigation Model (U3DSNM) ... 64
4.3 Technical Model: Python Classes ... 66
4.4 Map to IndoorGML and CityGML ... 66
4.5 Discussion ... 69
4.6 Summary ... 69

Chapter 5 Three New Path Options ... 71

5.1 Current Research on Navigation Path ... 71
5.2 Two New sI-space Related Navigation Path ... 72
5.2.1 Parameters ... 72
5.2.2 MTC-path ... 74
5.2.3 NSI-path ... 75
5.2.4 A Path Selection Strategy ... 76
5.2.5 Illustration of the Two Path Options ... 77
5.3 ITSP-path ... 80
5.3.1 Concepts and Modeling ... 82
5.3.2 Procedures of ITSP-path Planning ... 82
5.3.3 Illustration ... 83
5.4 Summary ... 85

Chapter 6 Reconstruction of 3D Navigation Spaces ... 87

6.1 Semi-indoor Space Reconstruction ... 87
6.1.1 Identification & Ordering of Proper Building Components ... 88

6.1.2 Determination of Top and Bottom & Space Generation 88
6.1.3 Space Trimming 90
6.1.4 Illustration 90
6.1.5 Algorithms 91
6.2 Semi-outdoor & Outdoor Reconstruction 94
6.2.1 Extract Object Footprints 94
6.2.2 Classify Semi-outdoor and Outdoor 95
6.2.3 Reconstruct 3D spaces 96
6.2.4 Illustration 96
6.2.5 Algorithms 97
6.3 Building Shells Reconstruction 100
6.3.1 Compute TIC by Projecting Footprints onto the Terrain 101
6.3.2 Set Height and Create Sides 101
6.3.3 Generate Top and Bottom to Reconstruct Building Shells 102
6.3.4 Rebuild Terrain Considering TIC as Constraints 102
6.3.5 Illustration 102
6.3.6 Algorithm 105
6.3.7 Other Possible Approaches of Building Shells Reconstruction 107
6.3.7.1 Footprints + Point Cloud 107
6.3.7.2 3D building model + DTM 108
6.3.7.3 3D building model + Point Cloud 110
6.3.7.4 Point Cloud 111
6.4 Summary 111

Chapter 7 Implementation & Case Study 115

7.1 Data, Software, and Flowchart for Implementation 115
7.2 Space Classification and Reconstruction 118
7.3 Space Selection and Navigation Network Derivation 122
7.4 Path Planning and Comparison of Results 124
7.4.1 Examples of Seamless Navigation 124
7.4.2 Example of MTC-path & NSI-path 126
7.4.3 Comparison of Results 129
7.5 Example of ITSP-path 130
7.5.1 Data Preprocessing 130
7.5.2 Navigation Network Derivation 131
7.5.3 ITSP-path Planning 135
7.6 Summary 139

Chapter 8 Conclusion and Recommendations 141

8.1 Conclusion 141
8.2 Conclusion on Topics 142
8.2.1 Environments 142
8.2.2 Spaces Representation 143
8.2.3 Unified Navigation Model and Path Options 143
8.3 Discussion 144
8.4 Recommendations for Further Research 146
8.4.1 Extend the Definition of Spaces 146
8.4.2 Space Subdivision Application 147
8.4.3 Include Obstacles in Path Planning 147
8.4.4 Evaluate the Navigation Performance 147
8.4.5 Extend the Results to Other Fields 148
8.4.6 Reconstruct Spaces Based on Other Data Source 148
8.4.7 Investigate Space Accessibility 149
8.4.8 Develop and Evaluate New Navigation Path Options 150

Papers Related to this Book **151**

References **153**

Index **171**

Preface

Contemporary living environments are getting more and more complex, combining structures of indoor and outdoor environments. These structures can be broadly subdivided into entirely enclosed or bounded (indoor), partially bounded (semi-bounded), and unbounded (outdoor). Normally, pedestrians would expect to be guided seamlessly when transferring from one kind of environment to another. Seamless here means human operations or interventions are not required when transitions happen between different navigation modes (e.g., driving and walking), which commonly takes place in different environments.

In the past decades, seamless navigation has gained much attention, and many approaches have been reported in the literature or made available as commercial applications. Most of them rely on connecting indoor navigation networks to outdoor road/street-based networks, by introducing anchor nodes. Yet, navigation paths from such approaches are not optimal, because the navigation networks in indoor and outdoor are created under different assumptions and rules. Furthermore, the semi-bounded structures such as gardens, balconies, sheds, and the spaces they delineate are often omitted, because their definitions and classifications are either missing or unclear.

This book presents novel ideas, concepts, models, and approaches to address comprehensively all kinds of environments and under unified creation rules. The presented ideas bring together the authors' research progress in this area over the past five years. The book addresses four major topics: (i) a generic space-based framework, which defines and parameterizes the environments for seamless indoor/outdoor navigation; (ii) a unified 3D space-based navigation model, which ensures a unified data structure, data management, and navigation network construction for all spaces; (iii) a set of approaches for automatically reconstructing 3D building shells, semi-indoor, semi-outdoor, and outdoor spaces, which allows to mimic the space-based approaches to derive a network widely adopted by indoor navigation approaches; and (iv) several new path semantics-based options, which make use of the semi-bounded spaces, such as the Most-Top-Covered path (MTC-path) and the path to the Nearest Semi-Indoor (NSI-path).

This book is intended to trigger thought and motivate further research and development. It can be used by students, researchers, and practitioners, who are working in this field. The concepts and the definitions are expected to provoke researchers from a range of fields. For example, the generic space-based framework may be interesting for research on urban heat islands or urban micro-climate, while 3D space modeling approaches can be utilized for urban sunshine analysis, flood disaster simulation, urban shadowing, or painting cost estimation. This book can also be used as introductory reading for those who want to engage in seamless indoor/outdoor navigation research and teaching.

About the Authors

Jinjin Yan is an Associate Professor at the Qingdao Innovation and Development Center, Harbin Engineering University, China. In October 2016, he started his PhD at the 3D geoinformation research group, Faculty of Architecture and the Built Environment, Delft University of Technology. In March 2018, he transferred to University of New South Wales, Australia. He received his PhD degree from the Faculty of Built Environment, UNSW, in November 2020. His research is on 3D modeling (indoor, outdoor, semi-indoor, and semi-outdoor), 3D space-based navigation, 3D analysis, 3D space subdivision, seamless indoor and outdoor navigation, BIM and GIS integration, Mobility as a Service (MaaS), and Super Navigation Network.

Sisi Zlatanova is a Professor at the Faculty of Built Environment, UNSW, Sydney, Australia and is leading the GRID Lab. She received her PhD degree from Graz University of Technology, Austria. She has worked as a software developer and has held academic positions at the University of Architecture and Civil Engineering, Sofia, Bulgaria, at Delft University of Technology, the Netherlands, and at Siberian State University of Geo-information Technology, Novosibirsk, Russia. She is the current president of ISPRS Technical Commission IV 'Spatial Information Science' and co-chair of the OGC SWG IndoorGML.

1 Introduction

Cities worldwide are expanding in size, height, and depth. Public buildings are growing more complex and incorporate a large variety of activities. Citizens can access the buildings on foot, by car, or using other wheeled devices. They can drive trailers or child buggies or can be in wheelchairs. Robots are increasingly in use for cleaning, guidance, or providing services. This complexity requires an understanding of the implications of building structures and their relation to the surrounding environment. New concepts to bridge indoor and outdoor developments are needed which would allow seamless navigation in indoor/outdoor environments for various modes of locomotion, including combinations of them. Seamless indoor/outdoor navigation has attracted attention from industry and academics in the past decades. Yet, the available applications are very simplistic, and the research is very sparse. Even the concept of navigation for a large part of people still consists of discrete, environment-dedicated applications.

This chapter introduces the concepts of navigation and summaries the current research on the seamless indoor/outdoor navigation. Then, concepts and terminology that are used in this book are introduced, which includes five existing terms and eleven novel terminologies (six for built structures, and five for space). The chapter concludes with an overview of this book, including the main contributions, structure, and brief introduction of chapters.

1.1 NAVIGATION CONCEPTS

Navigation (also called path-finding or way-finding elsewhere) is a fundamental activity in daily human life. It is one of the oldest research fields and focuses on monitoring and controlling the movement of a human or robot, using or not transportation means such as crafts or vehicles, from one place to another. Based on the places or areas where navigation is performed, navigation can be classified into four general categories: land navigation, marine navigation, air navigation, and space navigation. Often, navigation is related to the studies that deal with determining location and direction. Pedestrian navigation can be seen as part of land navigation as it reflects the case when a human is walking.

In research, pedestrian navigation is described as the method of determining the direction of a user from a familiar goal across unfamiliar terrain [1], or the process of orientation to reach a specific distant destination from the origin [2]. Navigation refers to three main questions: (location) where am I and where is the target? (path) where do I go? (guidance) how do I get there? Thus, navigation can be defined as the process or activity of accurately ascertaining locations of users, planning paths, and guiding them to follow the path to the desired destination. From a technological point of view, to be able to perform correct navigation, several components have to be available [3,4]: (3D) localization of start point and destination, a (3D) model that

DOI: 10.1201/9781003281146-1

represents the space subdivision and provides a network of connected spaces, (3D) algorithms for path computation (on a topological model or a grid), and guidance by text, voice, and visuals (on a 3D map or using augmented reality devices). The process of guidance may require identification of specific items (Points of Interest, Landmarks) as well as tracking of the user, alarming and correcting if the path is not followed.

People may refer to one or several of the above-mentioned components as "navigation". The term navigation is widely used for outdoor car guidance, and therefore some may consider only the global positioning systems such as GPS, Galileo, Baidu as "navigation". Outdoor navigation has been widely used in our daily life and diverse travel modes have been considered, such as walking [5], driving [6], transiting [7], and cycling [8]. It might not be immediately transparent that a navigation application, which is integrated in the positioning device, maintains a street network and calculates the shortest or fastest paths. Such positioning devices can also continuously register locations and create a "tacking path" of the moving person, without computing a path and maintaining a street network. Such paths also do not fall in the definition of navigation to be used in this book. In fact, as the name implies, such systems focus on positioning, which only has contributions on the first question of navigation.

Another commonly accepted use of the term 'navigation' comprises applications that can answer the three questions of navigation on the same platforms, such as Google Maps[1], OpenStreetMap[2]. Although such applications may require the user to provide their location due to a lack of positioning devices, they are typical examples of navigation systems.

Very often, the navigation is not intended for direct use, i.e., going from one point to other, but to investigate possible path options, or to estimate best path for a specific locomotion mode, or to evacuate people, or to avoid obstacles. In such cases, the human/vehicle/craft is replaced with an agent, and simulation and optimization are taking part, which needs agent/user requirements and/or preferences (i.e., agent profile), and information about the environments for navigation (spaces and their parameters, conditions, dynamic changes). Figure 1.1 illustrates this concept. For vehicle navigation, the agents/users are vehicles, navigation environments/spaces are generally outdoor, and navigation requirements could be the shortest/fastest path. In case of indoor navigation, the agents/users may be pedestrians, the environment is indoor, and the requirements may be "a path without climbing stairs".

With pedestrians moving seamlessly between one kind of environment to another (e.g., between buildings and surrounding areas), navigation guidance tools/systems should extend from merely outdoor or indoor guidance, to provide support in the combined indoor-outdoor context [9, 10]. That is, it is necessary to have seamless (integrated/continuous) navigation. Seamless navigation here is a term that describes universal navigation service, where human operations or interventions are not required when transitions happen, or the agents would not be aware of how the device

[1]`https://www.google.com/maps`
[2]`https://www.openstreetmap.org/`

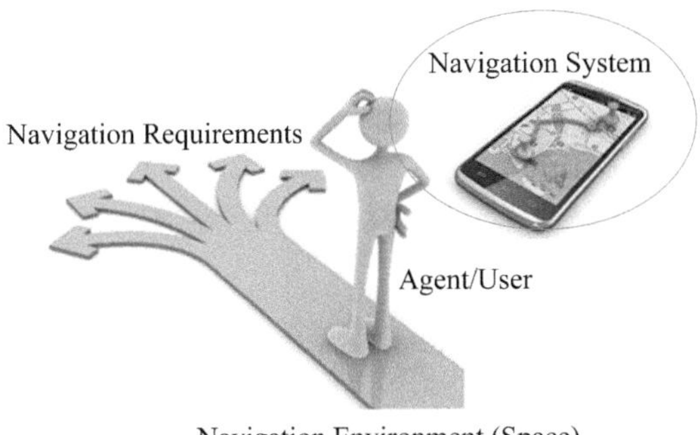

Figure 1.1 The components of a navigation system.

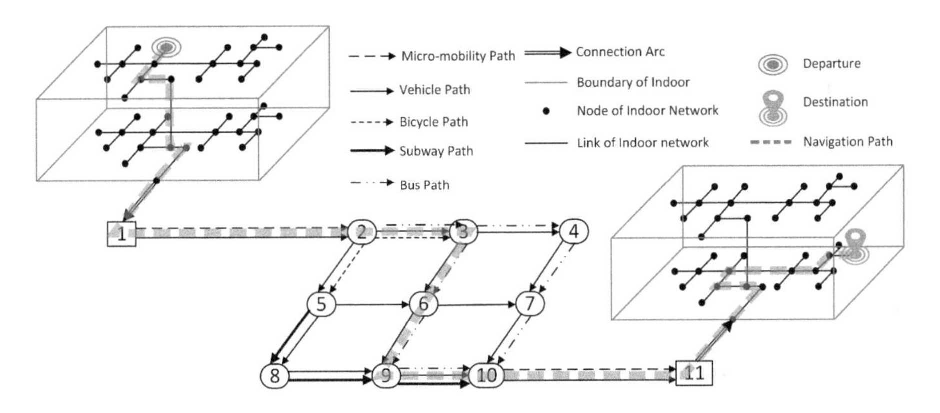

Figure 1.2 The ideal seamless navigation. No matter where a user is, he/she can get a navigation path from departure to destination directly, during which various transportation modes are involved.

works for the seamless services [11]. From a technical point of view, an application that provides seamless navigation should be able to provide indoor/outdoor localization, maintain an integrated indoor/outdoor model for navigation and provide means for guidance indoor and outdoor. Additionally, all possible spaces for navigation should be considered, i.e., indoor, outdoor, semi-indoor/outdoor. Very often semi-spaces are at the transition between indoor and outdoor, and cannot be omitted [12]. The ideal seamless navigation (tool/system) described by [10] can be seen in Figure 1.2.

1.2 CURRENT ATTEMPTS ON SEAMLESS NAVIGATION

In the past decades, seamless navigation has attracted a lot of attention and a tremendous number of approaches have been reported in the literature or made available as commercial applications in our daily life [13] for different travel modes (e.g., driving, walking, or flying). A successful seamless navigation should cover both indoor and outdoor environments [14] and consider how agents experience specific parts of the living environment when offering navigation services [10]. This section summarizes the existing two major approaches/attempts on seamless navigation in indoor and outdoor.

Several great attempts have been conducted to develop integrated indoor and outdoor navigation that provide seamless navigation path and guidance. Most current well-known efforts to combine indoor with outdoor navigation are focused on localization (tracking) techniques, in particular the transition of positioning systems from indoor to outdoor environments or vice versa [15]. Typically, outdoor localization is based on Global Navigation Satellite System (GNSS), while indoors exhibits much larger variety. There are some common indoor positioning technologies, e.g., Indoor Positioning System (IPS) [16], Inertial Navigation System (INS) [17], localization based on vision [18], Infrared [19], Ultrasonic [20], Radio Frequency Identification (RFID) [21, 22], Bluetooth beacons [23], Wi-Fi [24], Barcodes [25], and Quick Response (QR) codes [26–28]. For instance, the pedestrian navigation system developed by Hitachi[3] integrates wireless LAN and GPS. However, the integrated indoor and outdoor localization systems are important for positioning and tracking of users, but they do not have direct effects on path planning approaches. Therefore, this book will not focus on this approach.

Another approach of seamless navigation commonly used in research is to integrate indoor and outdoor navigation networks, i.e., developing a unified navigation network. The common practices are combining indoor navigation networks with outdoor road/street-based networks through the concept of anchor node [13, 29, 30]. Similar concepts named "entrance node" [31–33] are used by other researches to complete the same integration work (Figure 1.3). IndoorGML (this standard will be introduced in Section 3.2.1) and subsequent indoor/ outdoor seamless navigation research based on IndoorGML leverage anchor node to link the indoor navigation network with outdoor navigation network (Figures 1.3(a) and 1.3(b)). In addition, entrance node is another analogous structure for integration.

Such approaches assume that (i) the only missing part of constructing a unified navigation model is a linking strategy and (ii) the current indoor and outdoor navigation networks are sufficient for navigation applications in their environments. However, although the "unified" navigation model can provide a certain degree of improvement, they fall short in achieving seamless navigation.

Firstly, it is hard to exchange and maintain such integrated navigation networks, because the indoor navigation networks are generally extracted based on 3D (free) spaces, while the main source of navigation network in outdoor space is road/street networks. This has caused a lot of controversies, such as in pedestrian navigation, where a network might bring in inaccurate estimation of the length of the navigation path [10, 34].

[3] https://www.ogc.org/case_studies/hitachi

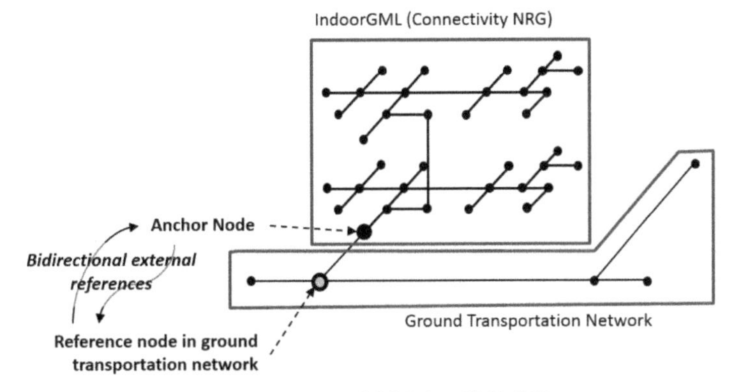

(a) Anchor node in OGC IndoorGML [30]

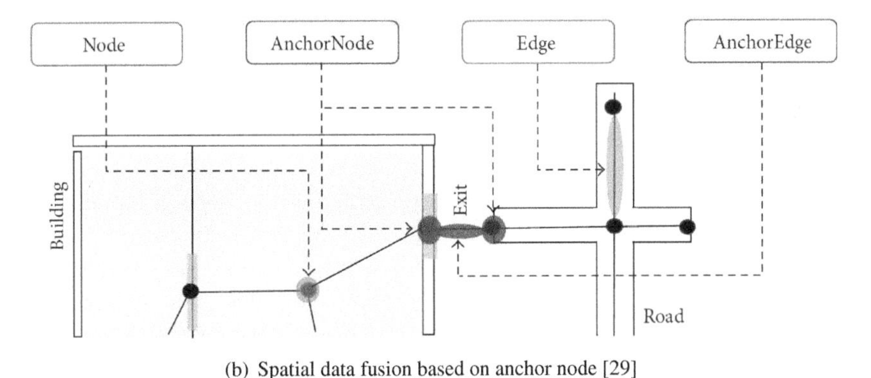

(b) Spatial data fusion based on anchor node [29]

Figure 1.3 Anchor node used for integrating indoor and outdoor navigation network.

Secondly, indoor and outdoor navigation networks have different resolutions, and they are generally dedicated to specific applications (indoor for pedestrians while outdoor for vehicles). For instance, a node in an indoor navigation network can be a room [32, 35] or functional areas [2, 36, 37], while a node in an outdoor navigation network is mostly related to items in transportation networks (intersection of a road). Thus, a pedestrian has to accept that some spaces (e.g., street elevator) will not be provided.

Thirdly, outdoor navigation networks are quite generalized. Different agents should have specific areas/spaces for their movements, except for the shared areas [38]. For example, the sidewalks of a road are restricted to pedestrians/wheeled users access only [5, 39], while the traffic lanes are dedicated to vehicles. If pedestrians are navigated based on streets, they might need to walk through unsafe areas [40–43].

Furthermore, some public spaces without clear paths (e.g., squares), semi-indoor (sI-spaces), and semi-outdoor (sO-spaces) are overlooked in navigation maps. Although more spaces are being considered in pedestrian navigation networks, including sidewalks, footbridges, undergrounds, etc. [44], their lack often leads to undesirable detours [40,45].

Therefore, it is still necessary to have a real unified navigation model that can overcome the drawbacks of the existing "unified" models and help with achieving seamless path planning in indoor and outdoor spaces.

1.3 NOTIONS AND TERMINOLOGY

This section introduces the concepts and terminology (five existing terms and eleven novel terminologies) that are used in this research. Using existing terms is the prior choice, but either no existing terms can precisely convey, or there are no appropriate existing terms to represent the information this book wants to present. For example, calling the structure that looks like a roof but only used for reinforcement (Figure 2.19) as a roof is inappropriate.

The first subsection presents five existing concepts employed in this book, in which agent is elaborated to clarify the recipients of navigation services. Other than that, footprints, building height, terrain, and terrain intersection curve (TIC), which are related to the 3D space reconstructions (Chapter 6). In the second subsection, six new terms for built structures are presented, which consists of top, side, bottom, topClosure, bottomClosure, sideClosure. The third subsection introduces five new terms for space. The eleven new terms are introduced as the basic terms of the generic space definition framework, attributes for the unified 3D space-based navigation model, and parameterized indicators of space size to help with selecting spaces for navigation.

1.3.1 FIVE EXISTING CONCEPTS

Agent Agents are users that use or engage in navigation activities or use resources offered by spaces. Agents can be broadly categorized as humans and robots. Further, based on their locomotion modes, they can be classified into walking (pedestrians), rolling (e.g., trolleys, wheelchairs), flying objects (e.g., drones) or combinations of these. In navigation, agents can vary in size and height and have certain requirements for spaces. Thus, agents should not be regarded as 0D points in the navigation, but 3D objects.

The path planning tests of this book are conducted based on a pedestrian. The definition of a pedestrian is a person traveling on foot, walking or running. Persons who travel on tiny wheels such as roller skates, skateboards, and scooters or on the wheels of self-balancing scooters are not considered as pedestrians because they have lost the flexibility of pedestrians. For instance, pedestrians can walk on stairs, but persons using (tiny) wheeled devices cannot unless they revert to traveling on foot.

A pedestrian is abstracted as a cuboid that has a length (l), width (w), and height (h) (Figure 1.4). According to the literature, moving people need extra buffer space around them [46], and they like to maintain a certain distance from walls and obstacles as a comfort distance [47]. The two kinds of buffer distances are denoted by l_{buf} and l_{com}, respectively. As seen in Figure 1.5, the slash-filled parts are physical

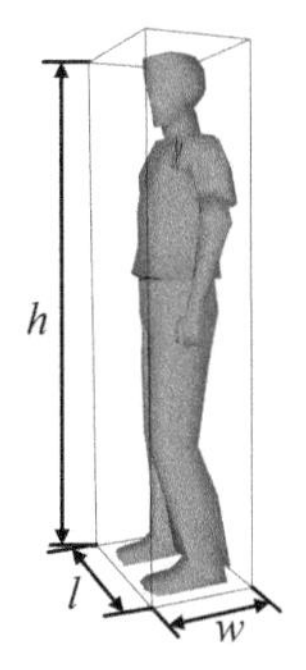

Figure 1.4 A pedestrian is modeled as a 3D object.

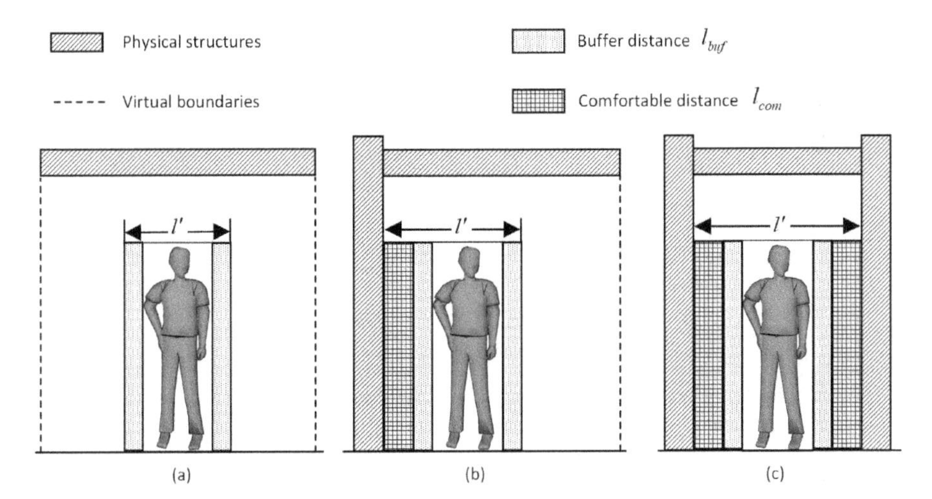

Figure 1.5 The size of the space required by a pedestrian. l', w', h' describe the size of space required by an agent whose size is l, w, h. The buffer and comfortable distances of width are not illustrated as the figures are side views.

structures, e.g., roofs and walls. The dashed lines indicate the virtual boundaries of the space. Dot-filled (l_{buf}) and grid-filled (l_{com}) parts are buffer and comfortable distances, respectively. Furthermore, taking the length direction as an example, space can be classified into three cases based on the number of sides, in which *(a)* has 0 side, *(b)* has 1 side, and *(c)* has 2 sides.

Thus, the minimum size of the space required by a pedestrian becomes $\{l', w', h'\}$, which can be computed by Equation 1.1.

$$\begin{cases} l' = l + 2l_{buf} + k_l * l_{com} \\ w' = w + 2w_{buf} + k_w * w_{com} \\ h' = h + h_{buf} \end{cases} \tag{1.1}$$

where l, w, h = size of a pedestrian

l	w	h	Description	Examples
625~875	375	> 2000	Stand uprightly	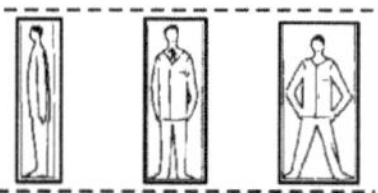
625~875	750	> 2000	Side by Side With back pack(s)	
800~2125	550~725	> 2000	With suitcase(s) or handbag	
750~2375	375~1125	> 2000	With walking stick or umbrella(s)	

Figure 1.6 Dimensions and space requirements based on body measurements (unit: mm). This figure is made based on the figures in Architects' Data [46].

l_b, w_b = buffer distances required by a moving pedestrian
l_c, w_c = comfortable distances required by a pedestrian to side(s) and/or obstacles
k_l, k_w = number of side(s) along the length and width ($k_{l,w} = \{0,1,2\}$)
h_{buf} = the buffer height required by a moving pedestrian

On the basis of the size of a pedestrian (Figure 1.6) and the two types of buffer distances for a moving pedestrian, the parameters for a pedestrian are shown in Equation 1.2. The extra buffer space is considered in this research is 10% of the pedestrian size [46]. The comfort distance is set as 10 cm, because [47] found that pedestrians like to maintain around 10 cm from walls and obstacles during walking.

$$\begin{cases} l, w, h = 625, 375, 1930 \\ l_{buf}, w_{buf} = 62.5, 37.5 \\ l_{com}, w_{com} = 100, 100 \\ h_{buf} = 19.30 \end{cases} \tag{1.2}$$

Then, considering that spaces generally are one or two side bounded, i.e., $k_{l,w} = \{1,2\}$, a pedestrian is modeled as a 3D object with length (750mm< l' <850mm), width (450mm< w' <650mm), and height (h' >1949.3mm) (Equation 1.3).

$$\{l', w', h'\} = \begin{cases} 850, 450, 1949.3 & k_l = 1 \ \& \ k_w = 0 \\ 950, 450, 1949.3 & k_l = 2 \ \& \ k_w = 0 \\ 750, 550, 1949.3 & k_w = 1 \ \& \ k_l = 0 \\ 750, 650, 1949.3 & k_w = 2 \ \& \ k_l = 0 \end{cases} \tag{1.3}$$

Footprints A footprint is the shape (e.g., 2D polygon) representing the projection of something (e.g., building) on a 2D plane (e.g., ground). The building footprint is

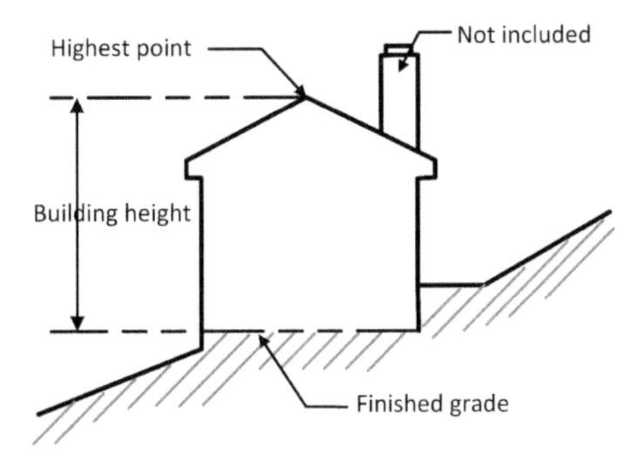

Figure 1.7 The building height defined by [55].

the area used by the building structure, which is defined by the perimeter of the building plan [48]. The building footprint is fundamental for a number of urban precinct modeling tools, such as the Envision Scenarios Planner [49]. Except buildings, city objects, such as streets and green areas, also have footprints [50].

Footprints of city objects can be found in a large number of different data sets, e.g., OpenStreetMap (OSM) [51], or reconstructed from airborne LiDAR data [52, 53], or extracted from digital surface models [54].

The footprints in this book are 2D polygons that represent projections of city objects, including buildings, shelters, roads, green areas, hand railings, enclosing walls, and fences.

Building Height Building height is a term to describe the vertical height of the building from the ground. The building height is a piece of critical information for 3D building shell model reconstruction based on the extrusion of footprints. Flat roofs or the accurate roofs may bring in different heights for a building, which will result in different building shell 3D models.

There are different ways to define the height of building, only the selection process of the starting point is slightly different. For instance, [55] defined the building height as the average maximum vertical height of a building or structure measured at a minimum of three equidistant points from finished grade to the highest point on the building or structure along each building elevation. Architectural elements such as parapet walls, chimneys, vents, and roof equipment are not considered as part of the height of a building or structure (Figure 1.7).

Another way of measuring the building height is based on the highest and lowest foundation points; if there is less than 3 meters elevation change between these two points, the height of the building is the distance from the highest foundation point to the highest point of the building; otherwise, the height becomes the distance between the lowest foundation point and the highest point[4]. For example, if the elevation changes between A and B for less than 3 meters (Figure 1.8), the height of the building is the distance from the highest foundation point (A) to the highest point of the building (D), otherwise the height becomes the distance between B and D.

[4]http://centralpt.com/upload/375/4785_Buidling\%20Height\%20Definition.pdf

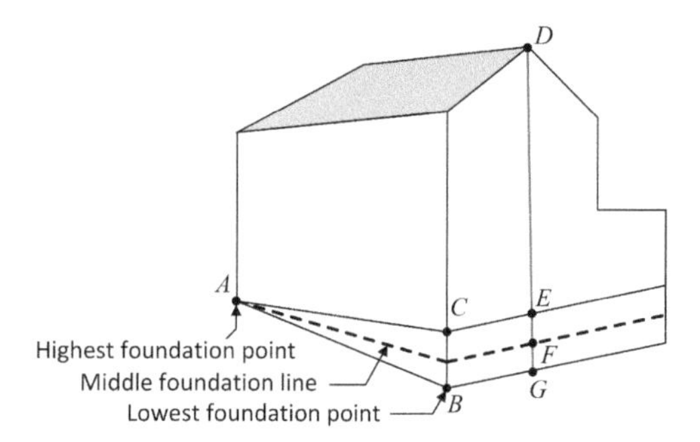

Figure 1.8 The rule of measuring building height. The *A* and *B* are the highest and lowest foundation points, respectively, *C* and *E* are on the same plane with *A*, *F* is the mid-point of foundation, and *D* is the highest point of the building.

This book considers the height of a building is the distance from the mid-point of the foundation to the highest point of the roof, no matter if the elevation change between the highest foundation point and lowest foundation point is more/less than 3 meters. For instance, the height of the building in Figure 1.8 is the distance between *F* and *D*. Therefore, the definition of building height is: the average maximum vertical height of a building or structure measured at a minimum of three equidistant points from mid-point of foundation to the highest point on the building or structure along each building elevation. Architectural elements of a building or structure, such as parapet walls, chimneys, vents, and roof equipments are not considered part of the height of a building or structure.

Terrain A terrain is referred to as ground or a tract of ground, especially with regard to its natural or topographical features or fitness for some use [56]. In the digital domain, the Earth's surface is commonly modeled by means of Digital Terrain Model (DTM) [57], such as Triangulated Irregular Network (TIN), which is an efficient alternative to the dense grid Digital Elevation Model (DEM) to present terrain surface [58]. It represents a surface formed of non-overlapping contiguous triangular facets that are with irregular sizes and shapes, which can describe in an intuitive way the continuous elevation changes of terrain. There are other alternative names, which also may represent different products, such as Digital Height Model (DHMs), Digital Ground Model (DGMs), as well as Digital Terrain Elevation Models (DTEMs). In this research, a terrain is a DTM that represented by Constrained Delaunay triangulation (CDT).

Terrain Intersection Curve (TIC) *TerrainIntersectionCurve(TIC)* is a curve that indicates, where 3D objects are touching the terrain (Figure 1.9), which is introduced by the CityGML to address the topological issue that the 3D objects may float over or sink into the terrain will occur when integrating 3D city objects with the terrain. That is, the terrain is locally fixed to fit the TIC. In this research, the TIC ensures correct positions and topological relationships between reconstructed 3D spaces and the DTM.

Figure 1.9 *TerrainIntersectionCurves* (TIC) in CityGML. The TIC is shown in bold black curve.

1.3.2 SIX NEW TERMS FOR BUILT STRUCTURES

Roof, wall, and floor/ground are three basic built elements, which play critical roles in forming and defining the structural boundaries for spaces. To allow more complex boundary configurations to be considered, three generalized notions (*Top*, *Side*, and *Bottom*) were introduced to replace the three building elements [12]. Their definitions are the following:

* **Top** represents a physical structure that can cover the space. A physical top may serve as protection from weather conditions (e.g., rain, wind, cold, heat) or as a limitation to estimate clearance (e.g., for flying or carrying large items). The top can be a human-made or natural structure/object (e.g., wood, stone, tree). The areas delineated with red rectangles in Figures 1.10(a) and 1.10(b) illustrate a roof and a top, respectively.
* **Side** indicates a physical structure that encloses a space from the surrounding directions. A physical side may act as protection from weather conditions (e.g., wind) or as a limitation to estimate entry possibilities (e.g., getting around to find a door, or flying above). *Side* is always consistent with the direction of the plumb line, otherwise, it is a *Top* or *Bottom*. The

(a) Roof

(b) Top

Figure 1.10 Example of roof and top (marked by red rectangles).

(a) Wall (b) Side

Figure 1.11 Example of wall and side (marked by blue rectangles).

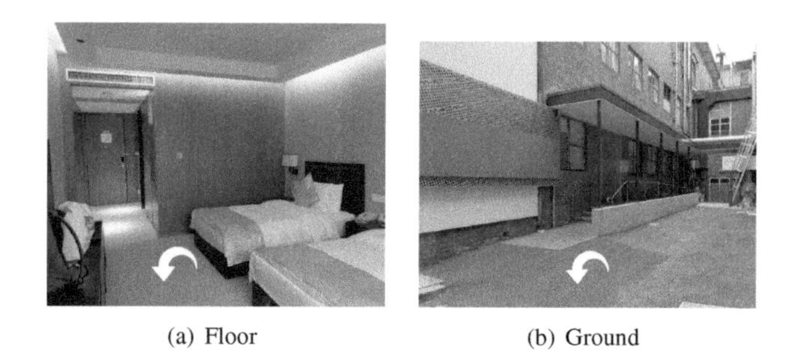

(a) Floor (b) Ground

Figure 1.12 Example of floor and ground (marked by white arrows). Both act as the bottoms of spaces.

blue rectangles in Figures 1.11(a) and 1.11(b) illustrate a wall and a side, respectively.

* **Bottom** A bottom is a structure that encloses a space from the lower direction and offers a platform where pedestrians can stand on it. Similar to the two former structures, a bottom can be an artificial (e.g., floor or slab) or a natural structure/object (e.g., ground). In this work, the ground/floor is assumed to be the default bottom structure and has always been assumed to be present. That is, space starts at the bottom. If there is no bottom, there is no space. The areas marked by white arrows in Figure 1.12(a) and 1.12(b) illustrate a floor and ground, respectively.

Closure To quantify the level of closure of the three structures, three other notions *topClosure*, *sideClosure*, and *bottomClosure* are introduced, which also have been presented in the authors' previous research [59]. The closures are three coefficients that have only numerical values. Their definitions are the following:

△ **topClosure** (C^T): the topClosure is a coefficient that expresses how much a space is physically bounded from its top. It corresponds to the ratio between the substantial (material) area and the entire area of the top boundary structure.

- △ **bottomClosure** (C^B): the bottomClosure is a coefficient that expresses how much a space is physically bounded on its bottom. It is then the ratio between the substantial (material) area and the entire side area of the bottom boundary structure.
- △ **sideClosure** (C^S): the sideClosure is a coefficient that expresses how much a space is physically bounded on its sides. It is defined as the ratio between the length of side parts physically enclosed by substantial (material) and the total side(s) length.

The C^T and C^B are similar, which can be computed by the Equation 1.4. The C^S can be computed by Equation 1.5.

$$C^{T/B} = a\_e/a\_t \tag{1.4}$$

where $C^{T/B}$ corresponds to either C^T or C^B, while a_e and a_t are the area of the enclosed part, and the total area of the top/bottom, respectively.

$$C^S = l\_e/l\_t \tag{1.5}$$

where C^S corresponds to sideClosure, while l_e and l_t are the length of physically enclosed part of surrounding boundary and the total length of surrounding boundaries, respectively.

Two simplified cases are employed to visually illustrate the computations of the C^T, C^B, and C^S (Figure 1.13). In the two cases, the slash-filled parts are physical structures, while the rests are empty.

The first case (Figure 1.13(a)) illustrates the closure computation of top/bottom, in which the area of the enclosed part is $a\_e = l_0 * W + l_1 * W + l_2 * W + l_3 * W$ while the total area of the top/bottom is $a\_t = L * W$, thus, the top/bottom closure of the case can be computed based on Equation 1.4, i.e., $C^{T/B} = a\_e/a\_t = (l_0 * W + l_1 * W + l_2 * W + l_3 * W)/(L * W)$. In comparison, the C^S of the case (Figure 1.13(b)) is

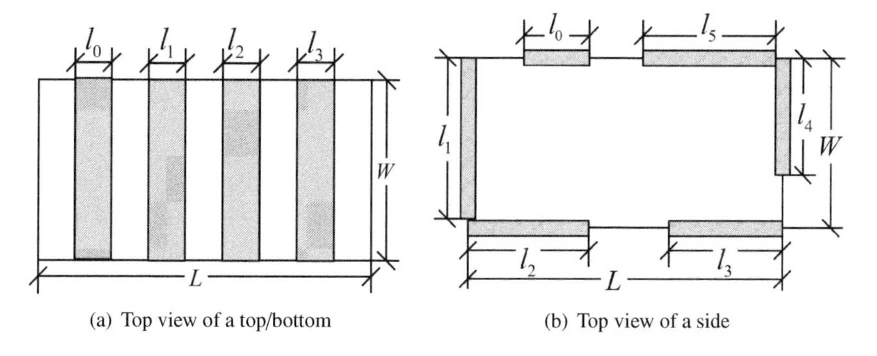

Figure 1.13 Illustration of closure computation, in which the slash-filled parts are physical structures. l_0 to l_5 are the length of the physically closed parts; L and W are the length and width of the top/bottom/side, respectively.

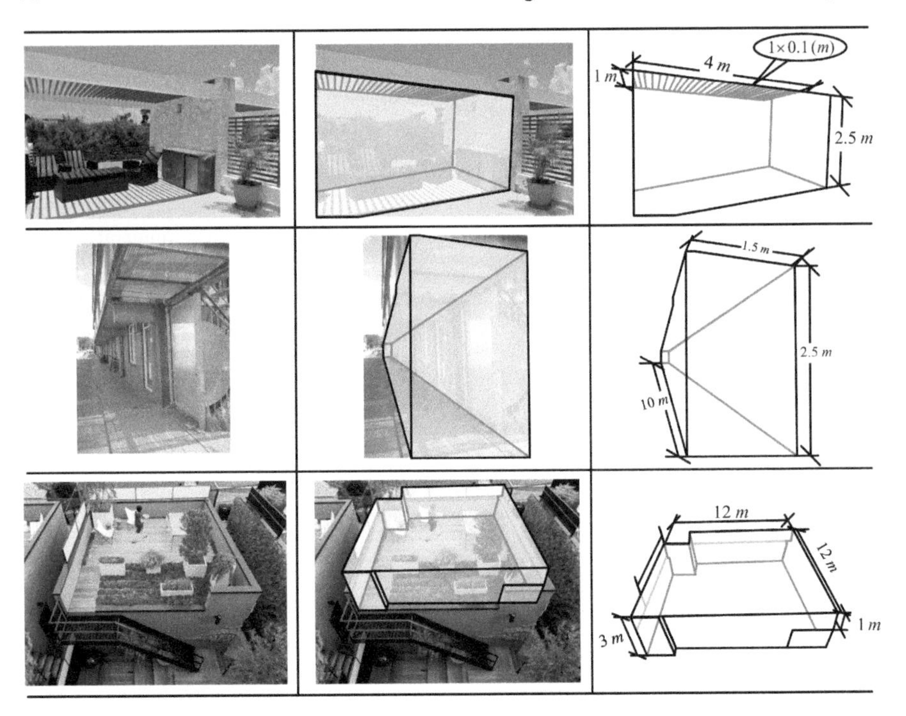

Figure 1.14 The abstraction of top, side, and bottom, and estimation of their closure.

$C^S = l\_e/l\_t = (l_0+l_1+l_2+l_3+l_4+l_5)/(2L+2W)$, in which thee length of physically enclosed part of surrounding boundary is $l\_e = l_0+l_1+l_2+l_3+l_4+l_5$ and the total length of surrounding boundaries is $l\_t = 2L+2W$.

Figure 1.14 illustrates how the top, side, and bottom are abstracted and how the C^T, C^S, and C^B are estimated for three real-built structures. The spaces are sketched as closed boxes. We estimate the area of each polygon of the box, compute the closure per polygon and average the closures when more than one polygon is involved.

The first case is a pergola, which has dimensions $4*1*2.5 = 10$. The total area a_t of the top is $4*1 = 4$ and a_e is 2.5 , because the top has 25 beams (each is $1*0.1$). Thus, $C^T = a\_e/a\_t = 2.5/4 = 0.625$. The space has four sides and one bottom, in which two sides are concrete walls (length of each is 1), and another two sides (each is 4) are entirely open. Therefore, $C^S = l\_e/l\_t = 2/10 = 0.2$, where $l\_e = 1*2 = 2$ and $l\_t = 1*2+4*2 = 10$. The bottom of this case is the part of the top projected vertically on the ground, thus, $C^B = 1$. The second case is an eave case, thus its $C^T = 1$. This space has one totally enclosed side (each is 10) and three open sides $(10+1.5*2)$. Therefore, the average of the four enclosures is computed and $C^S = 0.43$. Similar to the first case, the bottom of this case is the part of the top projected vertically on the ground, thus its $C^B = 1$. The third case is a roof terrace. The space has dimensions $(12*12*3)$. It has no top, hence its $C^T = 0$ and $C^S = 0.93$. The C^B of this case is also 1.0, but the bottom of this case is the whole roof.

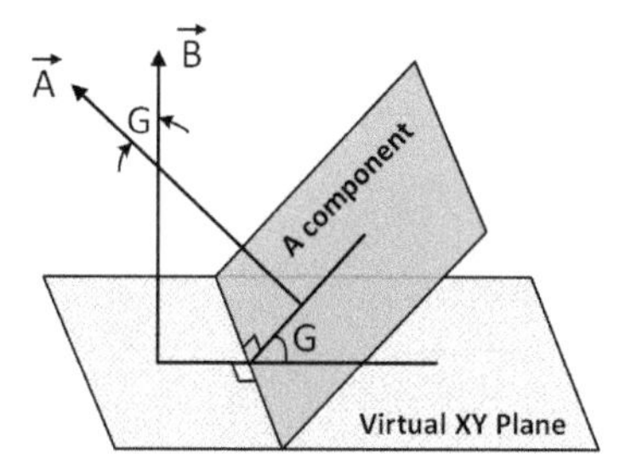

Figure 1.15 Illustration of the Gradient computation.

1.3.3 FIVE NEW TERMS FOR SPACE

Gradient Some built structures have slopes (e.g., inclined roofs/pavements to permit rainwater to run off), and these structures also may act as the top, side or bottom of some spaces. In other words, boundaries can be tilted at an angle. The gradient is marked as (**G**). This characteristic of a boundary is introduced to distinguish the side from top and bottom, specifically, if the **G** of a boundary is 90°, it is a side, otherwise, it is a top or bottom. Furthermore, this term can help to consider whether the slope of a bottom is moderate enough to ensure agents can have activities on it.

The **G** of a (polygon) component is referenced to a virtual XY plane. The normal vectors of a component and virtual XY plane are $\overrightarrow{A} = (a,b,c)$ and $\overrightarrow{B} = (0,0,d)$, respectively (Figure 1.15). Then, the gradient can be calculated by Equation 1.6, where the unit is degrees:

$$\mathbf{G} = arccos\left(\frac{|c \times d|}{d\sqrt{a^2+b^2+c^2}}\right) \times \frac{180}{\pi} \tag{1.6}$$

Boundary All the top/side/bottom are boundaries of spaces, but it is necessary to distinguish if a boundary is physical or virtual, because physical and virtual boundaries show different properties in navigation. Generally, the physical boundaries of a space are impassable for agents and virtual boundaries indicate the borderlines. In other words, by referring to this information, the entry possibilities to get into the space can be estimated for agents. If a boundary is virtual and shared by two space entities, the agents can navigate through this boundary to travel between the two spaces. The definitions of physical/virtual boundary are the following:

+ **Physical Boundary**: a physical boundary of a space is the boundary formed by a physical structure(s), for instance, wall, roof, floor, ground, even vegetation wall.
+ **Virtual Boundary**: a physical boundary of a space is the imaginary boundary that used to allow space enclosure.

The following illustrations intuitively show how the different boundaries are employed to picture semi-bounded spaces.

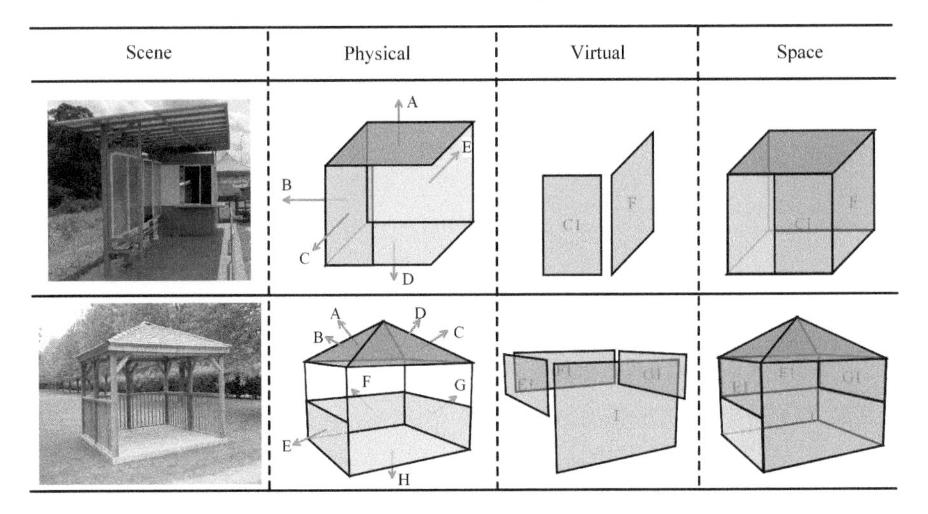

Figure 1.16 Two examples of semi-bounded spaces with tops.

There are two semi-bounded environments with tops: a bus stand and a gazebo (Figure 1.16). The hollow parts below their roof(s)/shelter(s) are semi-bounded spaces. The bus stand has five physical boundaries on the same plane: a quadrilateral top (*A*), three sides (*B*, *C*, and *E*), and a bottom (*D*). To make it as an enclosed 3D volume, two more missing virtual boundaries (*C*1 and *F*) are needed. Therefore, the boundaries *A*, *B*, *C*, *D*, and *E* are physical, while *C*1 and *F* are virtual. The gazebo has eight physical boundaries (*A* to *H*): *A* to *D* are four physical tilted triangular tops; *E*, *F*, and *G* are three physical sides; and *H* is the physical bottom. So, in this example, four virtual boundaries (*E*1, *F*1, *G*1, and *I*) are needed to make an enclosed space. The two scenes show that the shapes of boundaries are not limited to quadrilaterals and that they can be tilted.

Similarly, a yard is utilized to illustrate a semi-bounded environment without tops and how space is employed to picture it (Figure 1.17(a)). The hollow part within the fences and walls is the semi-bounded space. In Figure 1.17(b), *A*, *B*, *C*, *D*, and *F* are physical sides, and *E* is the bottom; *G* and *H* in Figure 1.17(c) are needed virtual top and sides when enclosing the space as an enclosed 3D volume. Thus, this semi-bounded environment is successfully pictured as a 3D space by physical and virtual boundaries (Figure 1.17(d)). It should be noted that, theoretically, the height of this semi-bounded space can be arbitrary, but in this illustration, the height is set as that of the highest wall(s).

Space Radius and Space Height The agent of this research is a pedestrian, who is modeled as a 3D object, thereby having certain requirements for spaces. Thus, considering if spaces are large enough to accommodate a pedestrian, the spaces employed for navigation should be selected based on their size.

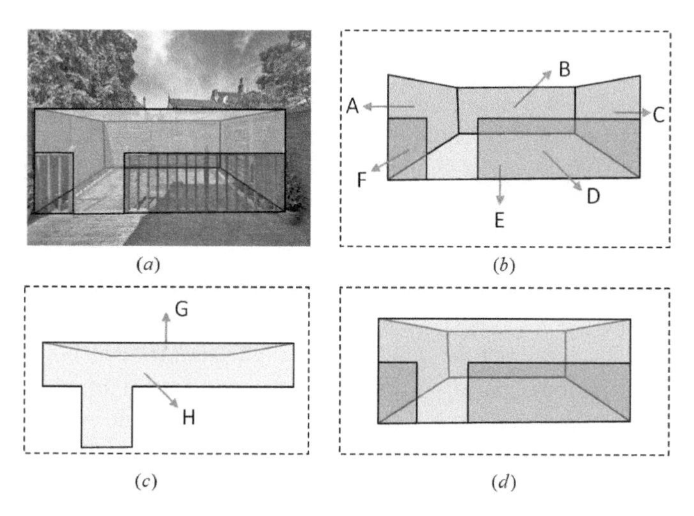

Figure 1.17 An example of a semi-bounded environment without tops: (a) yard; (b) physical boundaries; (c) virtual boundaries; (d) semi-bounded space.

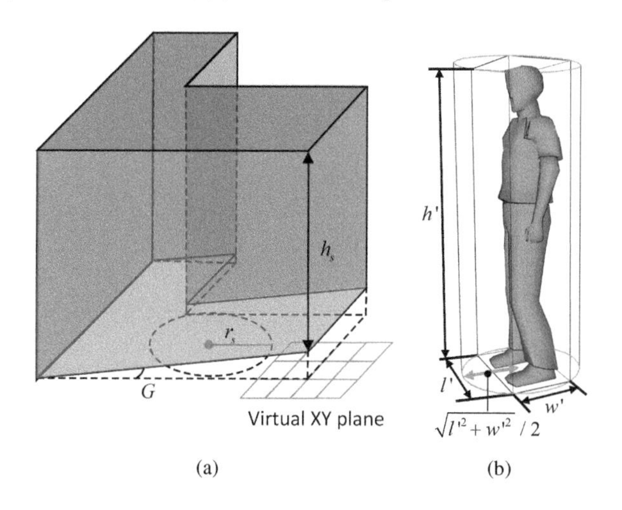

Figure 1.18 (a) Space radius (r_s), space height (h_s) and Gradient (G). Blue polygons are physical boundaries, while olive green polygons are virtual. (b) The minimum space required by a pedestrian. It is modeled as cylinder with bottom radius $\sqrt{l'^2 + w'^2}/2$ and height h'.

For the cuboid spaces, their size can be represented as a triple: {length, width, and height}. It is convenient for evaluating them by compare these three items with that of the minimum space required by pedestrian ($\{l', w', h'\}$, Equation 1.3), respectively. However, the spaces are not always cuboids, which makes this direct way not always valid. Therefore, Space Radius (r_s) and Space Height (h_s) are introduced (Figure 1.18(a)). Then, the both cuboid and non-cuboid spaces can be evaluated with the same way.

The definitions of the two terms are the following:

+ **Space Radius** (r_s): it is the radius of the maximum inscribed circle of the bottom projection. The computation process of the projection is projecting the bottom onto a virtual geometric *XY* plane along the *Z-direction*.
+ **Space Height** (h_s): it is regarded as the minimum height based on boundaries that touch the top and bottom at the same time.

To evaluate the non-cuboid spaces, the minimum space required by a pedestrian is further modeled as a cylinder, in which the bottom of the cylinder is the minimum circumscribed circle of the minimum required space (i.e., its bottom radius is $\sqrt{l'^2 + w'^2}/2$), and the height of the cylinder is equal to the height of the minimum required space (i.e., its height is h'), see Figure 1.18(b). Then, the evaluation process evolved into two steps: the first step is to compare the r_s with $\sqrt{l'^2 + w'^2}/2$, and the second step is to compare the h_s with h'. Only when spaces meet the following two conditions at the same time will be selected (Equation 1.7). Namely, (i) the space radius is larger or equal to the radius of the minimum circumscribed circle of the minimum space required by a pedestrian, and (ii) the space height is larger or equal to the height of the minimum space required by a pedestrian.

$$\begin{cases} r_s \geq \sqrt{l'^2 + w'^2}/2 \\ h_s \geq h' \end{cases} \tag{1.7}$$

where r_s is the space radius while h_s is the space height. l', w', h' are the length, width, and height of the minimum space required by a pedestrian, respectively.

1.4 BOOK OVERVIEW

This book consists of eight chapters:

Chapter 1 is the introduction of the book. It introduces the concepts of the navigation and the current research on the seamless indoor and outdoor navigation. Then, concepts and terminology that are used in this book are introduced, which includes five existing terms and eleven novel terminologies defined by this book. This chapter is ended with the overview of this book, including the main contributions, structure, and short descriptions of chapters.

Chapter 2 presents the contents of spaces that used for seamless navigation. After introducing the space definitions from literature review, a generic space definition framework based on authors' research is provided. The third subsection illustrates the thresholds and space classification based on the generic space definition framework with several real cases.

Chapter 3 presents the space-based navigation models. After briefly elaborating the motivation of employing navigation network model, this chapter focuses on the duality theory, 2D and 3D approaches for the network model derivation. Then, three international standards related to navigation (including IndoorGML, IFC/BIM, CityGML) are introduced. The final subsection presents an approach that can derive navigation network for QR code-based indoor navigation, which bridges the derivation approach to positioning techniques.

Chapter 4 provides the designed schema and class of a unified 3D space-based navigation model. Building upon investigations on the indoor navigation requirements, seamless navigation and spatial model, indoor/outdoor navigation service, requirements to a unified space-based navigation model are listed. The schema and classes make up the conceptual model. The second subsection shows a corresponding technical model by mapping its conceptual classes to python classes. Then, the Poincaré Duality theory that utilized as the basis theory of space-based navigation network derivation is introduced. Furthermore, taking QR codes as the positioning approach in indoor spaces as the example, the approach of deriving navigation network is illustrated.

Chapter 5 shows research on navigation path and presents three new navigation path options, including MTC-path, NSI-path, and ITSP-path. The three path options enrich the traditional shortest distance/time navigation path and open new directions for developing new path options. They can bring in a new navigation experience for pedestrians in indoor space.

Chapter 6 shows approaches of automatically reconstructing indoor, semi-indoor, semi-outdoor, and outdoor spaces as 3D spaces (volumes). On the basis the approaches, all four types of spaces can mimic the indoor environments to derive a network based on the 3D connectivity of spaces.

Chapter 7 implements the developed theories, approaches, and models, which include data preparation, space classification and reconstruction, space selection, unified space-based navigation model (navigation network) derivation, path planning, and comparison of results. Finally, the uncertainties and limitations of the whole work are discussed.

Chapter 8 concludes the whole book. Then, the recommendations on seamless navigation research, developments, and implementations are presented as further research topics.

2 Spaces for Seamless Navigation

In spatial science and urban applications, "space" is presented by multiple disciplines as a notion referencing our living environment. Space is used as a general term to help understand particular characteristics of the environment. In navigation, where agents move form one space to another, the notion of space is mainly related free of obstacles spaces, which are accessible/fit to navigate through. For instance, pedestrians can only walk in indoor spaces that are not occupied by furniture or building components (e.g., walls). Similarly, in outdoors, they can only walk in spaces that are not occupied by obstacles (such as railings, benches, trees, etc.). Despite the different aspects of indoor and outdoor environments (e.g., scale and dimension, landmarks, positioning technologies) [60, 61], spaces for indoor and outdoor navigation are conceptually equivalent. Hence, it is critical to formally define the space to support seamless navigation.

This chapter presents the characteristics of spaces that are used for seamless navigation. After reviewing several space definitions, a generic space framework is provided, which is furtehr detailed with introducing thresholds and space classification. Several real cases demonstrate teh use of teh framework.

2.1 SPACE DEFINITIONS

Space is an important notion in human lexicon aiming to indicate physical or imaginary parts of living environments. WordNet, one of the largest lexical English databases provides nine meanings of the noun "space", three of which are very relevant to this study [62]. Space is "the unlimited expanse in which everything is located", "an empty area, usually bounded in some way between things" and "an area reserved for some particular purpose". These three expressions clearly indicate the diversity in perceiving and describing space: empty or containing things, unlimited or bounded, physical or imaginary.

Most of the English dictionaries provide a more philosophical definition to reflect the fundamental importance of space to the understanding of the physical universe. Table 2.1 summaries the definitions found in seven dictionaries and lexical databases. As can be seen, four dictionaries consider space as continues", "boundless" and "empty". But space can also be perceived as portion of the space, such as space in a room. The definitions are not that explicit in specifying whether the spaces are empty or imaginary. The notion of limitless space is expressed by referring to its metrics in a three dimensional reference frame. The distinction between indoor and outdoor is introduced in a very intuitive way: the space inside of a structure (interior of a house, building, etc.) is indoor space, while the space outside the structure is outdoor space.

DOI: 10.1201/9781003281146-2

Table 2.1
The definitions of Space, Indoor, and Outdoor from different resources.

Sources	Space	Indoor	Outdoor
Princeton WorldNetWeb [63]	Space is either unlimited expand, or an empty area usually bounded in some way between things (e.g., the architect left space in front of the building), or an area reserved for some particular purpose (e.g., the laboratory's floor space).	Indoor is located, suited for, or taking place within a building, e.g., indoor activities for a rainy day; an indoor pool.	Outdoor is also called as out-of-door or outside, which is located, suited for, or taking place in the open air, i.e., outdoor clothes; badminton and other outdoor games; a beautiful outdoor setting for the wedding.
Oxford Dictionary [64]	Space is a continuous area or expanse which is free, available, or unoccupied; An area of land which is not occupied by buildings.	Indoor is situated, conducted, or used within a building or under cover. The origin of indoor is from in (as a preposition) + door in early 18th century, superseding earlier within-door.	Outdoor is a done, situated, or used out of doors.
Merriam-Webster [65]	Space is a boundless three-dimensional extent in which objects and events occur and have relative position and direction, e.g., infinite space and time. Or a physical space independent of what occupies is called also absolute space.	Indoor is of or relating to the interior of a building; done, living, located, or used inside a building.	Outdoor is of or relating to the outdoors; not enclosed: having no roof. done, used, or located outside a building.
Dictionary [66]	Space is the unlimited or incalculably great three-dimensional realm or expanse in which all material objects are located and all events occur. It also can be defined as the portion or extent of this in a given instance; extent or room in three dimensions, for instance, the space occupied by a body.	Indoor means occurring, used, etc., in a house or building, rather than out of doors, e.g., indoor games.	Outdoor means taking place, existing, or intended for use in the open air, e.g., outdoor games, outdoor clothes. Also out-of-door; out of doors; in the open air; the world outside of or away from houses.

(*Continued on next page*)

Table 2.1
(*Continued*)

Sources	Space	Indoor	Outdoor
Cambridge Dictionary [67]	Space is an empty area that is available to be used. From the dimension of view, the space is the area around everything that exists, continuing in all directions.	Indoor means happening, used, or existing inside a building, e.g., indoor sports/activities, an indoor racetrack/swimming pool.	Outdoor means existing, happening, or done outside, rather than inside a building, e.g., an outdoor swimming pool/festival, outdoor clothes; Or liking/relating to outdoor activities, such as walking and climbing.
Collins Dictionary [56]	Space refers to an area that is empty or available, and the area can be any size. Space means the unlimited three-dimensional expanse in which all material objects are located. Space is the whole area within which everything exists.	Indoor means situated in, or appropriate to the inside of a house or other building; Or of the inside of a house or building; Or living, belonging, or carried on within a house or building.	Outdoor, also out-of-door, means taking place, existing, or intended for use in the open air, e.g., outdoor games, outdoor clothes.
TheFree-Dictionary [48]	Space is an infinite extension of the three-dimensional region in which all matter exists.	Indoor is situated in, or appropriate to the inside of a house or other building, e.g., an indoor tennis court; indoor amusements.	Outdoor means the open air. Or an area away from human settlements.

In daily life, humans more often refer to portions of space such as areas or locations rather than universal unlimited space. Everyday expression such as "the space is packed with people", "there is no space in this room", "the kitchen space is spacious" suggest that the people tend to think of spaces as singletons (unique and self contained) enclosed by physical or imaginary boundaries. The singletons are assigned a broad spectrum of properties, and these properties can range from personal to communal. Ashihara [68] argues that we can distinguish between bounded space and unbounded nature space, in which bounded space is considered to be positive space, since it is created to fulfill human (who use this space) intentions and functions while the unbounded nature space is negative. This book embraces this human reference of space.

This approach is preserved when space is modeled by different disciplines in digital counterparts of real world. Researchers and developers discretize space into portions to be able to introduce useful properties, represent relations and visualize them [3]. Examples of such properties are weather conditions (temperature,

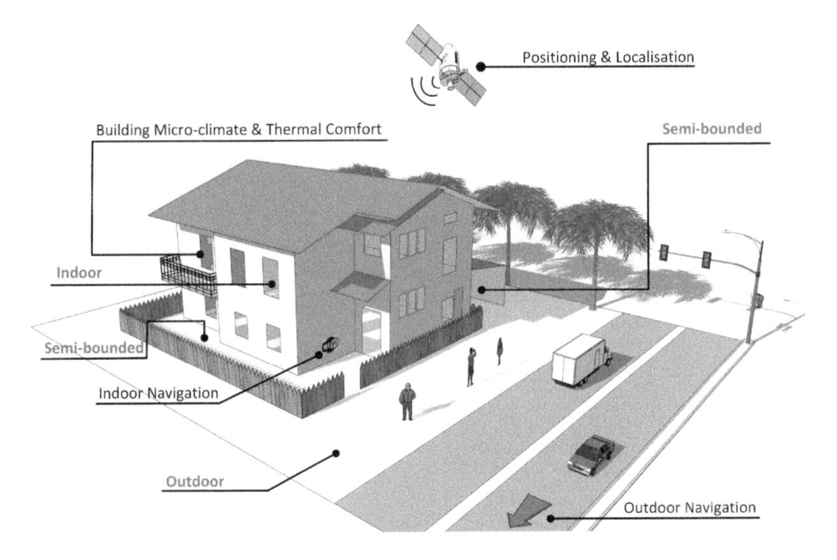

Figure 2.1 The domains in spatial science and urban applications that introduced space.

humidity, wind), accessibility (accessible, partially accessible, non-accessible) or legal rights (right to cross, ownership). As the literature shows the portion of space has been referred to as to space unit, space cell or as just space. For the scope of this book we will use the more formal notion of *Cell* when referring to a bounded portion of space.

Barriers such as walls, floors, ceilings are commonly used to create or separate indoor cells. Imaginary boundaries can be used to distinguish between space units outdoors for landscape planning. Combination of imaginary and physical boundaries can be used to partition spaces for leisure, drying laundry, growing plants [69], sheltering from the sun and wind [70] and similar activities [71–73]. Cells have been introduced in different disciplines related to urban applications (Figure 2.1), such as positioning and navigation [4, 12, 59, 61, 74–82] (indoor [2, 31, 37, 83–87], outdoor [14, 60, 88]), building micro-climate and thermal comfort [69, 89–94], landscape, urban planning & design [95–100], urban heat island [101–103], interior design [104–106], transportation [107–110] and intelligent space [111–114].

However, disciplines may compose cells differently for the same environment, which can raise issues in referencing or addressing spaces. This book is the first summaries the similarities, overlaps and differences while referring to spaces or portions of spaces in different fields. Therefore, we present the space concepts developed within the individual disciplines first and later we analyses and compare the finding. It should be noted that this book concentrates on space definitions in the context of built environments and geographical space for sake of seamless navigation. Therefore, treatment of space in philosophy, mathematics, physics, cosmology, psychology and social sciences is out of the scope of this book.

(a) Entirely bounded (indoor)

(b) Unbounded (outdoor)

Figure 2.2 Living environments in our life.

2.2 SPACE CLASSIFICATION

2.2.1 LIVING ENVIRONMENTS

With the expanding of cities worldwide, we can find that the buildings are expanding in size, height, and depth, which makes contemporary living environments more and more complex. Generally, environments entirely physically bounded by building components (e.g., walls, floors, roofs) are named as indoor [3, 14, 60, 85, 115] (Figure 2.2(a)). In contrast, the unbounded, connected and unoccupied environments are commonly perceived as outdoor (Figure 2.2(b)). Objects on the ground can provide some indications of boundaries of outdoor environments, but many of them would remain unbounded in the vertical direction, such as streets, pavements, squares, rivers.

Beyond that, there are many semi-bounded living environments that are difficult to be classified as indoor or outdoor sharply. They have characteristics of both indoor and outdoor, but cannot be seen clearly as either of them. In the literature, the most important structural feature of a semi-bounded environment is the upper boundary (e.g., roof, shelter). Therefore, we can temporarily distinguish the semi-bounded environments based on if they have upper boundaries.

We can find a lot of semi-bounded environments with upper boundaries formed by built structures in our living environments. Examples of such cases are overpasses, shelters for bus stops, gas stations, porches, courtyards and so on (Figure 2.3). Figures 2.3(a) and 2.3(b) are two semi-bounded environments with upper boundaries, but the former has surrounding boundaries, and the latter has not. A variety of structures can act as the source of upper boundaries, e.g., indoor environment (Figure 2.3(c)), bridge (Figure 2.3(d)), or shelter (Figure 2.3(e)). Furthermore, the function of upper boundaries can be different; some of them can help agents escape from rain or strong sun (Figures 2.3(a), 2.3(b), 2.3(c), 2.3(d), and 2.3(e)), but some cannot (Figure 2.3(f)).

We also can easily find many semi-bounded cases without upper boundaries (Figure 2.4). This kind of environment is without the upper boundary but bounded

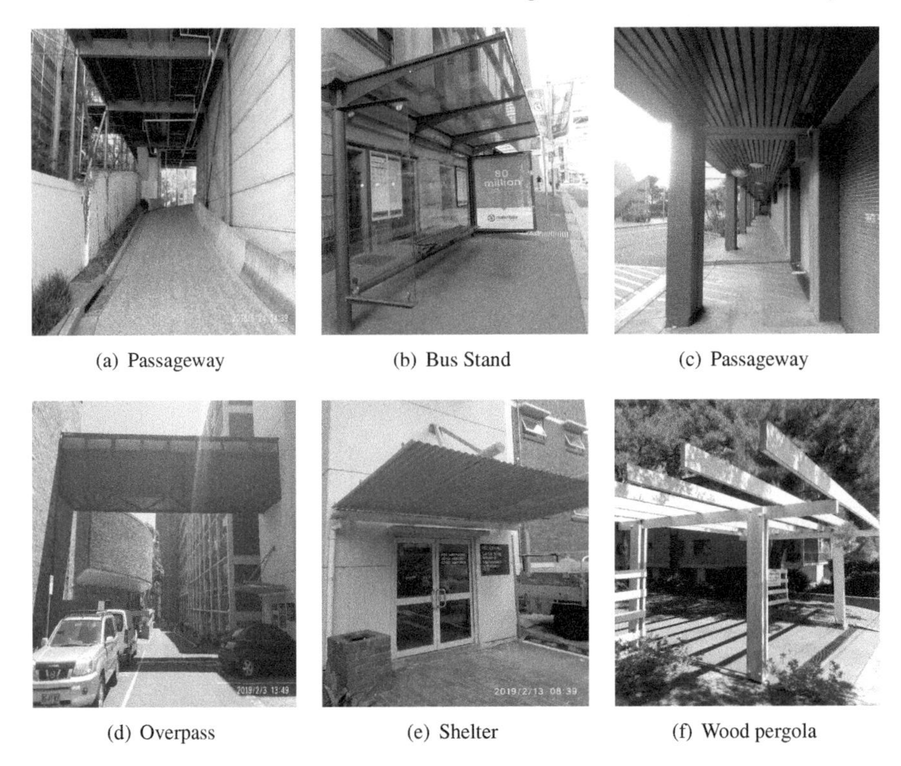

(a) Passageway (b) Bus Stand (c) Passageway

(d) Overpass (e) Shelter (f) Wood pergola

Figure 2.3 Examples of semi-bounded environments with upper boundaries.

or enclosed on the surrounding direction(s). On the point of upper boundaries, this kind of environment is making it similar to outdoor. A variety of structures can act as the source of surrounding boundaries, e.g., fence (Figures 2.4(a), 2.4(b), and 2.4(c)), building (Figure 2.4(d)), walls (Figure 2.4(e)), or combination of walls and fence (Figure 2.4(f)). Surrounding boundaries can indicate boundaries of a space, which is useful for space modeling.

2.2.2 INDOOR & OUTDOOR

Spaces in navigation have been classified as being located either indoors or outdoors. No strict definition for outdoor is found in the literature, except that people regard objects in the open air as outdoor (assuming unbounded from above), such as streets, pavements, squares, rivers. In contrast, research dealing with indoor environments is more explicit with partitioning spaces into cells and attempts to provide formal definitions. For instance, IndoorGML [30] defined the space as to be bounded by architectural components, which is similar to the definition that indoor space is a building environment (such as a house, or a commercial shopping center), where people usually behave in [115]. Zlatanova et al. [3, 85] defined the indoor space as a place bounded by physical boundaries (e.g., walls, floor, doors) and intended to

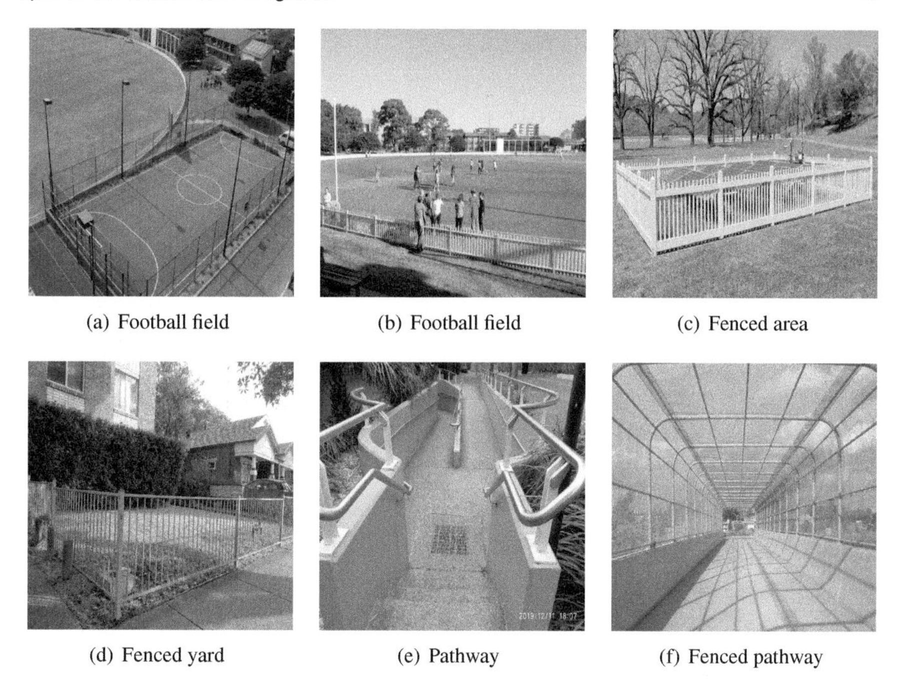

(a) Football field (b) Football field (c) Fenced area

(d) Fenced yard (e) Pathway (f) Fenced pathway

Figure 2.4 Examples of semi-bounded environments without upper boundaries.

support human activities. Indoor space is often referred to as to a physically enclosed space. Underground enclosures, which offer platforms for human activities, are also referred to as indoor space [60]. Winter [116] presented an indoor space definition with the analogy to the human body. The body is a container bordered by the skin. Similarly, the wall, floor, roof, fence can be seen as the "skin". According to these definitions, semi-enclosed spaces such as a veranda or an inner court would be outdoor spaces. Indoor space in [14] refers to a built environment rather than natural. According to these authors, underground cavities (caves, natural passages) would be classified as outdoor. [117, 118] investigated way-finding in the public transportation infrastructure based on traffic networks, in which the authors suggested to classify the environments for navigation into two types: network space and scene space. Network space consists of the public transport network. Scene space consists of the environment at the nodes of the public transport system, through which travelers enter and leave the system and in which they change means of transport. That is, outdoor spaces covered by navigation networks are network space while the spaces between indoor and outdoor are the scene space.

2.2.3 SEMI-BOUNDED SPACES

Only a few papers discuss semi-enclosed spaces that cannot be clearly attributed to indoor or outdoor. Examples are covered footbridges, sheds, balconies or partially roofed courtyards. Winter [116] defined these spaces as transition zones by giving examples and further proposed that ubiquitous navigation must be able to deal with the contrasting properties and conceptualizations of outdoor and indoor environments, and with the spaces in between. [80] defined these spaces as transitional spaces. There can be also extended environments, potentially vaguely bounded, where a crisp distinction between indoors and outdoors is difficult. More examples are tunnels, enclosed footbridges, or partially roofed courtyards.

The key point of distinguish a semi-bounded space is the upper boundary, which makes semi-bounded spaces are mostly mentioned in research related to building micro-climates, because the upper boundaries can have a significant influence on the climate of the space. Many papers have discussed architectural designs to improve residential and building micro-climates and reduce cooling and heating energy requirements [69,70,121,129]. Typically, three main spaces are identified: semi-indoor, semi-outdoor, and connection/transition/buffer and an attempt is made to provide definitions (Table 2.2).

Table 2.2

Classifications and definitions of spaces between indoor and outdoor in the field of building micro-climate and thermal comfort.

Name	Brief definition	Examples	References
Semi-indoor space	A space covered with canopies that is related to a building, and can combine indoor and outdoor climate conditions.	Balcony (installed with external windows), courtyard, open air kitchen with 3 walls, open space equipped with overhead shed/arcades.	[69, 71, 93, 94, 119, 120]
Semi-outdoor/ semi-open space/ semi-open space	A space that is not enclosed entirely, and has some settings including human-made structures that moderate the effects of the outdoor conditions	Space with eave, courtyard, sheltered space, sheltered balconies, outdoor space partly enclosed by a semi-transparent pitched roof	[70, 89, 121–126]
Connection/ transition zones/ buffer areas/ transitional spaces	Spaces that are located between indoor and outdoor but neither indoor nor outdoor.	Tunnel-like underpass, tunnels, enclosed footbridges, partially roofed courtyards	[80, 116, 127, 128]

The authors of [119] defined the covered space as a semi-indoor space, which is partially surrounded by indoor spaces. [90] presented that a semi-indoor space can be created by using a special roof (Vela Roof) as cover to passively avoid uncomfortable (coldest and overheated) conditions, thereby reducing the energy demand significantly. Moreover, they defined space that is not entirely enclosed by walls, windows, doors, etc. as outdoor. The semi-indoor space defined in the research [91] is a semi-indoor stadium, which has a roof that can be used to close the indoor volume to a relatively large extent. Thus, spectators and equipment are protected from wind, rain, and snow. This space still has direct openings to the outside. In contrast, [92] took the stadium as a semi-outdoor space in the assessment research of thermal comfort. In the research of condensation in residential buildings [69], semi-indoor spaces are created by installing external windows to balconies in Korean apartment units, which are used as environmental buffer spaces to improve comfort and reduce cooling and heating costs. In the research of the impact of improved cook-stoves on indoor air quality in the Bundelkhand region in India, [120] mentioned the kitchen with 3 walls is semi-indoor compared with outdoor (open-air) kitchen with a makeshift thatched roof for summer. [93] showed the open space equipped with overhead sheds is semi-indoor spaces, which can provide the citizens with sheltered space for public activities. In their research, they argued that illuminating such a huge semi-indoor space only by artificial lighting is against the energy-saving principle. Thus, they added lighting ducts to enable natural light to travel through the plate of the collector shed and reach the hall on the ground floor. In South Korea, traditional markets have been enclosed as semi-indoor by installing arcades along street edges to improve their physical environment [71], e.g., to alleviate inconveniences caused by inclement weather. [94] considered semi-indoor and semi-outdoor spaces are two transitional spaces for thermal comfort. In their examples, a semi-outdoor space is covered by a fabric membrane while semi-indoor space is a studio of 8m high where its roof has 33% of zenital apertures for natural lightning. In contrast, [130] took the space, which is partially open towards the outdoor environment as a semi-outdoor space. They even reinforced the concept that outdoor space partly enclosed by a semi-transparent pitched roof (e.g., glass roof) is a semi-outdoor space in a later research [124]. [131] defined the semi-outdoor as locations that, "while still being exposed to the outdoor environment in most respects, include human-made structures that moderate the effects of the outdoor conditions." Examples include roofs acting as radiation shields or walls acting as vertical windbreaks. [89] defined the semi-outdoors as exterior spaces that are sheltered and attached to the building. The authors also mentioned that an outdoor environment indicates a space without any covering to provide shelter and an indoor space refers to a naturally ventilated room. The micro-climate of the semi-outdoor (partially enclosed space) usually has a lower effect of wind and is less hot than the outdoor [124]. Semi-outdoor spaces in [132] are areas covered by large roofs, leaving a direct connection with the outdoor environment. Museums and cultural center gardens, university campuses, shopping and leisure areas, hotels and resorts, are a few examples of building environments where covered semi-outdoor spaces are commonly integrated. In [133], semi-outdoor spaces are defined as the spaces which are partly open in the direction of the outdoor.

Three categories are introduced: inside the buildings such as entry atrium; covered spaces; shaded spaces, situated in an outdoor environment entirely. Covered streets are regarded in this category. Furthermore, semi-enclosed space [70, 125], or semi-open space [126] are used to name the space that is not enclosed entirely, and has some settings including human-made structures that moderate the effects of the outdoor conditions.

Some research offered the definitions only by examples. For instance, the bus shelter is a semi-outdoor in [134], because it can offer shelter in the form of a roof. The semi-outdoor space in [135] refers to the internal architectural space with maximum exposure to the lobbies, corridors, atrium, courtyards, passages, and verandas. In the research of building micro-climate and summer thermal comfort in free-running buildings with diverse spaces, [121] named the space between indoor and outdoor as semi-outdoor space with the example of the space combination of eaves section and courtyard. Similarly, authors of [89, 122] defined semi-outdoor as exterior spaces that are sheltered and attached to the building. Balconies are regarded as shaded semi-outdoor spaces to provide the much needed thermal relief to the occupants of flats during the hot seasons [123]. [136] defined indoor and semi-indoor spaces as GPS-denied environments.

2.2.4 FOUR EXAMPLES OF SPACE CLASSIFICATION AND DEFINITION FRAMEWORK

In past research, a few space classification and definition framework based on certain rules are introduced, such as spaces are defined and classified based on the buildings, the sensor (e.g., GNSS) reception perspective, or the types of the signal received by the users, or if a space is fully enclosed.

A typical example is the space classifications provided by [78]. The authors used lightweight sensing services to analyze indoor/outdoor environments with respect to positioning options for mobile applications. They partitioned and classified space into indoor, semi-outdoor and outdoor (Figure 2.5). The outside of a building is defined as outdoor, while the inside is indoor. Close to or semi-open building space is considered as semi-outdoor.

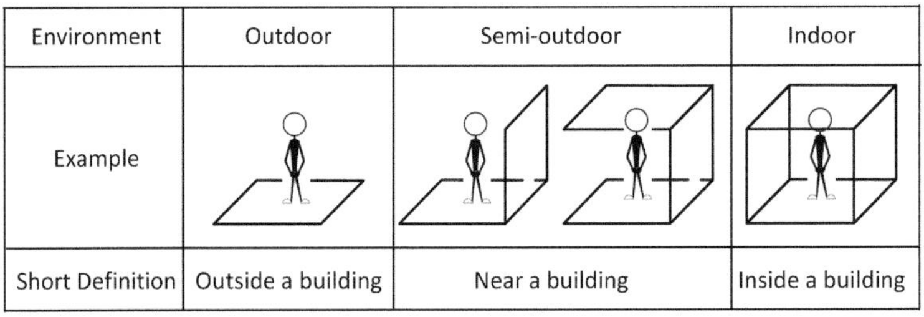

Environment	Outdoor	Semi-outdoor	Indoor
Example			
Short Definition	Outside a building	Near a building	Inside a building

Figure 2.5 Space classification based on building.

Environment	Open Outdoors	Semi-Outdoors	Light Indoors	Deep Indoors
Definition	Outside a building	Near a building	In a room with windows	In a room without windows
Example				

Figure 2.6 Space definition according to the reception of satellites signal [81].

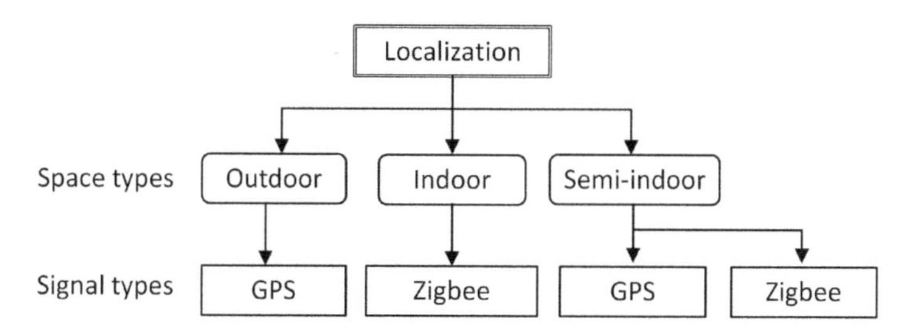

Figure 2.7 Space classification based on the type of signal received by the users.

The research [81] also provides a space classification according to the reception of the GPS signal as follows: open outdoors, semi-outdoors, light indoors, and deep indoors. As shown in Figure 2.6, areas which have an open sky condition (i.e., unbounded from above) and provide enough satellites for positioning are open outdoors. Areas such as urban canyon or wooded area, are classified as semi-outdoors. Light indoor is similar to semi-outdoor but inside the building. These are areas around windows, which still have some satellite availability. Deep indoors refer to places without any satellite coverage.

The research [137] developed a visual aid to the visually impaired person, in which they classified the navigation environments into three types, indoor, outdoor, and semi-indoor. The criterion for this space partitioning is the type of signal received by the users. Specifically, outdoor is the space which only can receive GPS signal, indoor is where ZigBee signal (communication protocol) is dominant, and semi-indoor is the portion of space where both signals can be detected (Figure 2.7).

Based on if a space is (fully) enclosed, the authors of [138] suggested classifying and defining spaces into indoor, quasi-indoors, quasi-outdoors, and outdoor, but no strict definition is provided. For example, a courtyard surrounded by buildings is quasi-indoors, because the yard is an inseparable part of the surrounding building and should be included as part of the building's indoor map to ensure continuity in navigation.

2.3 SPACE REPRESENTATION

Space in digital world is an abstract expression of the specific properties of environment and therefore is discretized into cells. To be able to manage, analyses

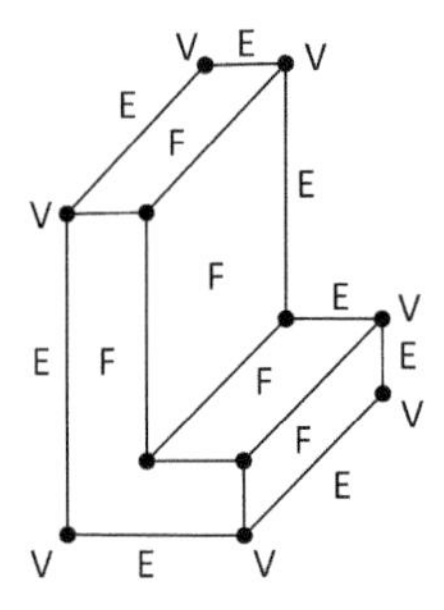

Figure 2.8 An example of BReps, in which a solid is represented by F faces, E edges, and V vertices.

and visualize the space, geometric representations are applied. For instance, in research such as positioning and navigation, building micro-climate and thermal comfort and landscape, spaces are abstracted and represented using Boundary Representation (BRep) [139], Constructive Solid Geometry (CSG) [140] or Spatial Occupancy Enumeration [141]. While appropriate for realistic visualization, Boundary representation can fall short in performing spatial operations as volume validation and computation. Therefore, for many applications Spatial Occupancy Enumeration (e.g., voxels [142, 143]) can be seen as alternative [144, 145].

2.3.1 BREPS

A boundary representation (BRep) is a geometric and topological description of the boundary of an object. The object boundary is segmented into a finite number of bounded subsets, called faces. A face is represented in a BRep by its bounding edges and vertices (e.g., Figure 2.8). Thus, a 3D BRep consists of three primitive entities: faces (2-dimensional entities), edges (1-dimensional entities) and vertices (0-dimensional entities) [146].

Spaces in indoor navigation are generally modeled as 3D volumes by BReps, such as multilayered space-event model [147], and 3D object based navigation [148].

2.3.2 VOXELS

Another 3D space representation is using voxels (one kind of Spatial Occupancy Enumeration) (Figure 2.9). Space is modeled by a set of cells with the same size marked as obstacle or non-obstacle.

Based on this representation, navigation paths can be computed by checking the availability of cells with respect to their neighbors. This representation fills out the indoor environment with obstacle and non-obstacle voxels (usually represented as cubes). The obstacle voxels can further be classified according to the needs of the application and can be static, semi-mobile and mobile. Path computation using this representation depends on the voxel size. If the voxels are too coarse, important information or space might be lost, while fine voxels can increase the time for processing

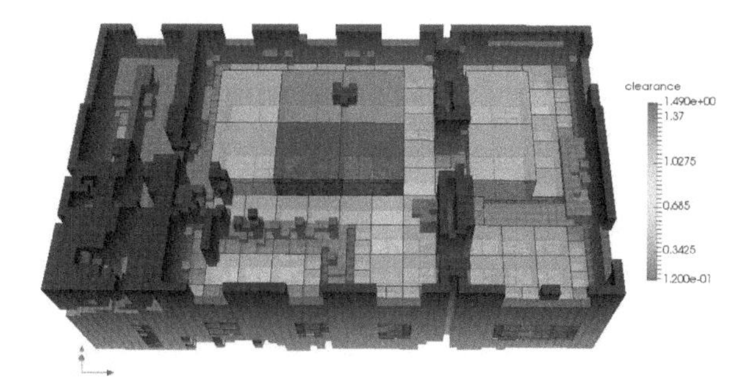

Figure 2.9 All spaces including walls are represented by voxels and organized in an octree structure [149].

and need more memory to be stored, which could be problematic for outdoor navigation. To reduce the storage size and improve the performance, data structures as octree are largely utilized (Figure 2.9).

2.3.3 EXAMPLES OF SPACE GEOMETRIC REPRESENTATIONS

There is no strict pattern, according to which geometric representation is employed in different disciplines. It many cases, the geometry used depends on the geometric expressions of the modeling software, i.e., GIS (Geographic Information Systems), CAD (Computer-aided Design) or BIM (Building Information Modeling). Although, spaces are often visualized, there are cases where "space" is just a perception. For example, intelligent transportation spaces are given as abstract concepts and we have not found evidences that they have been explicitly represented with geometry.

Spaces in indoor navigation are generally modeled as 3D volumes. There is variation on the use of term space in navigation research. While some research tends to related air and obstacles as to space, there are definitions according to which space is only the 3D hollow part bounded by physical or imaginary (non-existing) elements [84]. The international OGC standard IndoorGML, which provides formalism for a space-based navigation model, [30] relates to a space as being "navigable" (i.e., hollow) or non-navigable (e.g., furniture, walls) (Figure 2.10). It further maintains different space subdivisions in a multilayered space-event model [147]. This approach allows to consider 3D object-based navigation [148]. 3D navigable spaces are used to derive a navigation network applying Poincaré duality theory [150]. Nodes in the network are associated with spaces units, which can also represent landmarks or decision points, and edges between them represent the connectivity between spaces. Moving from one node to another is allowed only when there is an edge between them. Commonly cost of edges indicates distance or travel time between nodes [151] and nodes can contain semantic information about the location (name, type, description, etc.).

This research [152] proposed an approach to extract 3D indoor spaces in complex building environments based on indoor space boundary calculation, which includes

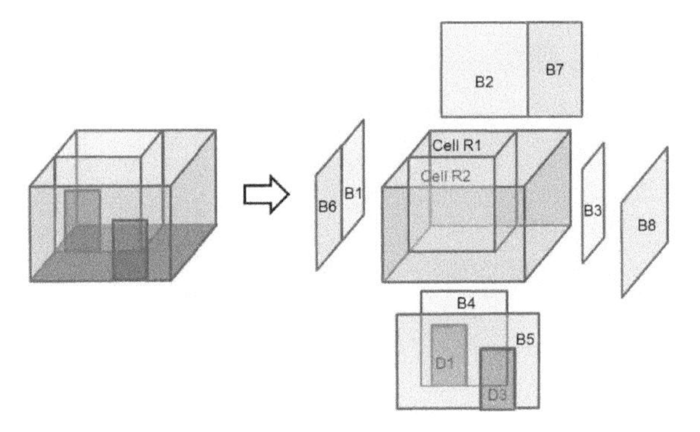

Figure 2.10 Example of the indoor 3D space modeling in IndoorGML [30].

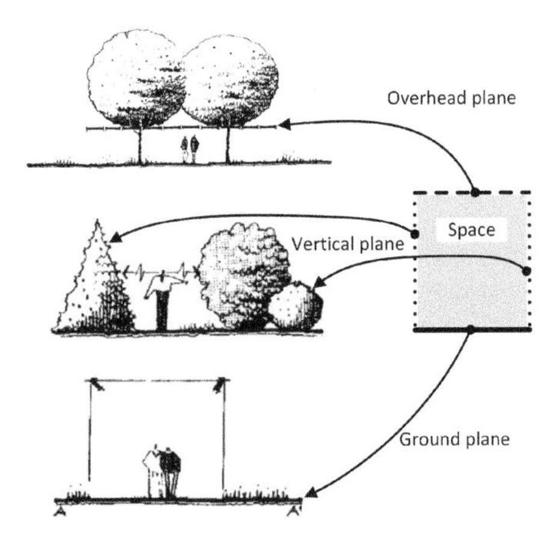

Figure 2.11 Example of geometric modeling of space in the landscape. Adapted from the figures in [96].

three steps: the Boolean difference for single-floor space extraction; relationship reconstruction; and cross-floor space extraction. Landscape spaces are also represented by geometric representations, in which boundaries are constructed by surrounding planes, including ground plane, vertical plane, and overhead plane [96]. As seen in Figure 2.11, treetops of arbors act as overhead planes, bushes as vertical planes, and ground/terrain as the ground plane. This modeling approach is similar to the method of building space modeling. It means that the ground plane, vertical plane and overhead plane can correspond to *Bottom*, *Side*, and *Top*, respectively.

In the filed of building micro-climate and thermal comfort, indoor, outdoor, semi-indoor/semi-outdoor notions are widely used to represent spaces, but the geometric representation is not that explicit. This can be partly explained by the fact that, on one hand, definitions of these spaces are notion-based or example-based only, and on the other hand, factors (e.g., temperature, humidity, wind speed, and solar intensity) are the main characteristics. Yet, considering the spaces are the similar to those used in navigation, we deem it possible to apply the same modeling approach as in navigation.

Almost in all outdoor related research fields (such as landscape, urban planning & design, urban heat island, transportation), outdoor spaces are represented as 2D surfaces as they are naturally unbounded in the Z direction. Only few research papers considered outdoor spaces as 3D and applied 3D geometric approaches. [153, 154] reports a model to subdivide continuous urban spaces into convex and solid voids (3D volumes) for analysis and classification based on physical measures, in which outdoor is roughly classified as urban void and the subdivision procedure is conducted based on convex polygon. [155] proposes a versatile data model for analyzing urban architectural void, in which space compartmentalization is conducted based on Gestalt theory and its model can be rendered both as a 2D and 3D representation.

2.4 A GENERIC SPACES DEFINITION FRAMEWORK

Current research on space classifications and definitions are mainly notion-based or example-based. This section introduces a generic space definition framework, which brings in systematic and parametric definitions for all types of spaces, where the navigation activities may happen. The definitions can support the development of algorithms to enclose semi-bounded and entirely open spaces. As such, this framework ensures a uniform parameterized space definition that enables the inclusion of all types of spaces into the navigation model for seamless path planning. The generic space definition framework that formally define and classify all the spaces where the navigation activities happen as the union of four types of spaces: indoor (I-space), outdoor (O-space), semi-Indoor (sI-space) and semi-Outdoor (sO-space), as expressed in Equation 2.1. It is necessary to point out that: (i) this classification means the semi-bounded spaces are classified and defined as semi-indoor and semi-outdoor; (ii) any space for navigation must belong to one of the four defined categories; and (iii) all the four types of spaces are completely bounded by boundaries and separated from each other. They may share a point, edge, or face (boundary), but they do not have any overlaps between each other. Their formal definitions are shown in the following sections.

$$S = I \bigcup O \bigcup sI \bigcup sO \tag{2.1}$$

where S, I, O, sI, and sO denote Space, I-space, O-space, sI-space, and sO-space, respectively.

2.4.1 DESCRIPTIVE DEFINITION

The characteristics of different environments (spaces) (Table 2.3) give hints to distinguish them from the structures: Top, Side, Bottom, and entirely bounded. The definitions of Top, Side, Bottom are introduced in Section 1.3.2. On the basis of

Table 2.3
Characteristics of living environments (spaces), in which "✓" denotes a space has the structure while "×" denotes has not.

Environments(Spaces)	Top	Side	Bottom	Entirely bounded
Indoor	✓	✓	✓	✓
Outdoor	×	×	✓	×
Semi-indoor	✓	✓/×	✓	×
Semi-outdoor	×	✓	✓	×

their characteristics, indoor and outdoor spaces are easily identified, because an indoor space is entirely bounded by top, side, and bottom while an outdoor space has bottom only.

Semi-bounded environments are further categorized as semi-indoor and semi-outdoor, i.e., the semi-bounded environments with upper boundaries are classified and defined as semi-indoor spaces, while that without upper boundaries as semi-outdoor. The reason why these two notions (semi-indoor, semi-outdoor) are employed is the semi-indoor space is more similar to indoor while semi-outdoor is similar to outdoor. In particular, sI-spaces have tops and that have a significant influence on the climate of the space. sO-spaces are open from the upper direction like outdoor spaces. The biggest difference between them is the former is partly or entirely enclosed from surrounding directions, but the latter is entirely unbounded. Thus, during navigation, agents like pedestrians/vehicles can visit an outdoor space from all surrounding directions, while they have to go through specific structures (door/entrance or similar) to visit a semi-outdoor space.

Then, the descriptive definitions of the four categories of spaces are the following:

+ **Indoor (I-space)** is the spaces that physically and entirely enclosed by tops, sides and bottoms.
+ **Semi-indoor (sI-space)** is the spaces that are semi-open to the outdoors, physically enclosed by top(s) in the upper direction, and may have a side(s) but is not physically enclosed completely like indoor. The bottom is assumed to be present by default.
+ **Semi-outdoor (sO-space)** is the spaces that are open to the outdoor from the top direction but enclosed by physical sides from surrounding directions.
+ **Outdoor (O-space)** is the spaces in the outdoor that completely open from the top and surrounding directions.

2.4.2 QUANTITATIVE DEFINITIONS

The descriptive definitions only provide impressions about space classification, which still cannot define spaces precisely and quantitatively. For instance, based on the descriptive definitions, both the spaces, bus stand (Figure 2.3(b)) and wood pergola (Figure 2.3(f)) should be classified as sI-spaces, however, they have a very big difference when focusing on their tops. The top of the former is a shelter that can be used for sun and rain protection, while that of the latter has only decorative and structural reinforcement function. For navigation use, it is more reasonable to classify the bus stand as an sI-space while the wood pergola as an sO-space or O-space. Therefore, this research further provides the quantitative definitions based on top closure (C^T) and side closure (C^S).

This section introduces five thresholds: $\alpha, \beta, \gamma, \delta$, and η to provide an estimate of C^T and C^S of a space, in which α, β and η are dedicated for C^T while γ and δ for C^S. By controlling these thresholds, the spaces are defined and classified quantitatively. Figure 2.12 offers a visual clue on how these thresholds define the boundaries and consequently the different spaces. On the two axes, the values of C^T and C^S are represented, where 0 means no closure and 1 means total closure.

Thresholds for C^T There are three threshold values related to C^T: α, β and η

- η allows to distinguish between indoor and outdoor spaces. A space with a $C^T \geq \eta$ is either an I-space or a sI-space. Otherwise, it is an O-space, or a sO-space.
- α is a closure value of the O-spaces. Specifically, a space with a $C^T \geq \alpha$ cannot be an O-space. It must be a sO-space, I-space, or sI-space.
- β is the lower threshold of the I-spaces. Thus, a space with $C^T < \beta$ cannot be an I-space. I must be a sI-space, sO-space, or O-space.

η illustrates how critical the notion of C^T is in determining whether space is closer to the indoor or the outdoor space, see Figure 2.12(a). Because the notion of indoor is hardly feasible without a minimum of a top bounding element to space, as reflected in the literature. But different constraints apply to the C^S, see Figure 2.12(b).

Thresholds for C^S There two threshold values related to C^S: γ and δ.

- γ is the upper threshold for the O-spaces with respect to C^S. A space for which $C^S \geq \gamma$ cannot be an O-space. It must be sO-space, sI-space or I-space.
- Similarly, δ is the lower threshold of the I-spaces such that a space with a $C^S \leq \delta$ cannot be an I-space. It must be sI-space, sO-space or O-space.

Unlike the C^T, C^S gives the meaning of the boundary between the indoor and the outdoor, but it is less critical in our framework. It provides estimates for the limits of the I-space and O-space but does not affect sO-space and sI-space. For this reason, only two thresholds are introduced and must be combined with the three thresholds of the C^T. Using the closure coefficients and their ranges, the four space types can be formally defined.

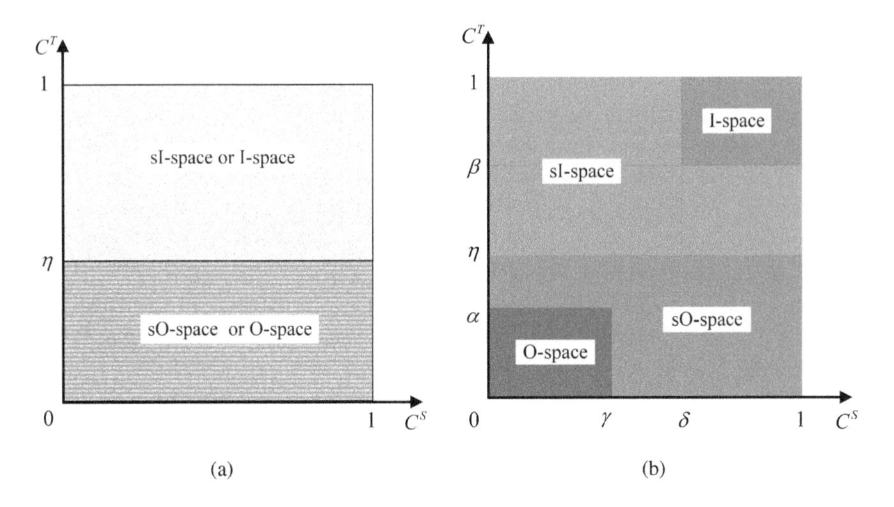

Figure 2.12 The C^T and C^S of different spaces. C^T and C^S are two physical structure parameters. α, β, γ, δ, and η are five thresholds. (a) sharply classify space into two categories, sI-/I-space, and sO-/O-space, by using η of C^T; (b) classify sI-/I-space into I-space and sI-space, and sO-/O-space into sO-space and O-space, by employing other four thresholds (α, β, γ, and δ).

Quantitative Definitions A space $S(C^S, C^T)$ can be defined as being one of the four types of space, as follows (Equation 2.2):

$$S(C^S, C^T) = \begin{cases} O & \text{if } C^S < \gamma \text{ and } C^T < \alpha \\ sO & \text{if } C^S \geq \gamma \text{ and } C^T < \alpha \\ & \text{or } \alpha \leq C^T \leq \eta \\ sI & \text{if } C^S \leq \delta \text{ and } C^T > \eta \\ & \text{or } \eta \leq C^T < \beta \\ I & \text{if } C^S \geq \delta \text{ and } C^T \geq \beta \end{cases} \tag{2.2}$$

Figure 2.13 illustrates the process of space identification. As mentioned above, C^T is critical to distinguish between the two indoor and two outdoor spaces. More top closure intuitively leads from outdoor to indoor. Outdoor space is actually never physically bounded but rather bounded by structures of the other subspaces surrounding it (sO-space, sI-space or I-space). Figure 2.14 shows sketches of possible cases along with real scene examples to illustrate an application of the definition framework on spaces relevant to navigation.

2.4.3 ILLUSTRATION OF THE GENERIC SPACE DEFINITION FRAMEWORK

To provide a sense of the types of closure and help in understanding the generic space definition framework, this section uses several simplified cases to illustrate the thresholds of C^T and C^S and shows the classification results.

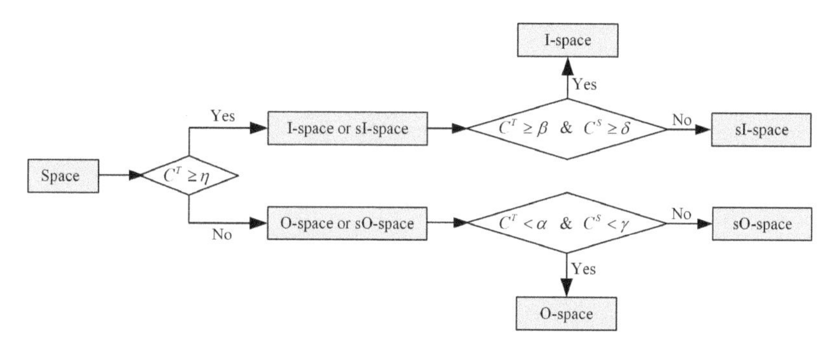

Figure 2.13 Flowchart of the space definition framework.

Thresholds of C^T For the visually sensing the thresholds of C^T, a simple rectangular structure is employed, which provides four example closures, see Figure 2.15.

The size of the example structure in Figure 2.15 is $2 * 1$(m), i.e., $a\_t = 2\ m^2$. a_e has different values in the four cases. In case (a), a_e=0.1 m^2 and therefore $C^T = 0.05$. Similarly, the closure of the remaining cases can be computed. a_e (gray parts) of case (b), (c), and (d) are 1.2 m^2, 1.9 m^2, and 2 m^2, their C^T are 0.6, 0.95, and 1, respectively. Visually, the first two cases (a) and (b) give the impression of being not well-sealed, while the second two cases (c) and (d) look like better sealed. On the basis of this visual inspection, this research sets the values of the three thresholds (α, β, and η) for C^T in Figure 2.12 as: $\alpha = 0.05, \beta = 0.95$, and $\eta = 0.6$.

Thresholds of C^S Based on occurrence of physical surrounding boundaries, the sO-spaces and O-spaces can be simplified into six cases (Figure 2.16), in which the case with two surrounding boundaries have two different variations (Figure 2.16(c) and (d)). Based on the Equation 1.5, the C^S of the six cases can be computed, i.e., $C^S_a = 0$, $C^S_b = 0.25$, $C^S_c = C^S_d = 0.5$, $C^S_e = 0.75$, and $C^S_f = 1$.

Environment	I-space	sI-space	sO-space	O-space
Definition	$C^T \geq \beta$ & $C^S \geq \delta$	$C^T \geq \eta$ & $C^S \in [0,1]$ *except* $C^T \geq \beta$ & $C^S \geq \delta$	$C^T < \eta$ & $C^S \in [0,1]$ *except* $C^T < \alpha$ & $C^S < \gamma$	$C^T < \alpha$ & $C^S < \gamma$
Example				
Scene				
Thresholds	$0 \leq \alpha \leq \eta \leq \beta \leq 1$ & $0 \leq \gamma \leq \delta \leq 1$			

Figure 2.14 Definitions of the spaces based on the definition framework.

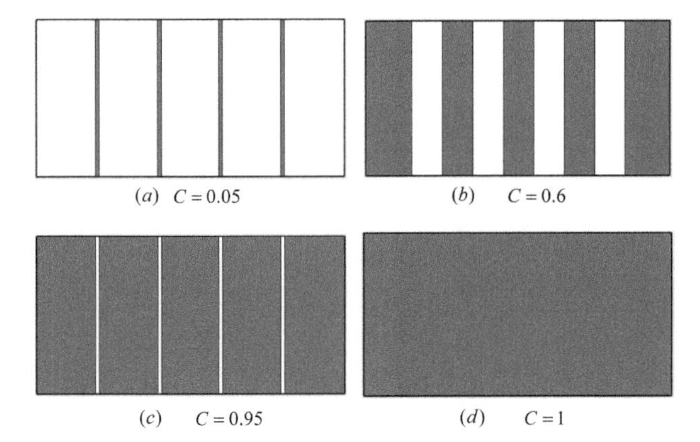

Figure 2.15 The structure with four different closure cases, in which the gray parts are enclosed (a_e), while white parts are open. Size of the example structure (a_t) is $2*1(\mathrm{m}^2)$.

The space differences can be clearly seen by visual inspection, however, it is still challenging to classify them into sO-space and O-space quantitatively. Hence, for the presentation purpose, this research sets the threshold of C^S on the basis of the visiting pattern of sO-space (go through a door/entrance or a similar structure when the agent is a pedestrian or vehicle), i.e., $\gamma = 0.75$ and $\delta = 0.95$. If a space without upper boundary structure, but its $C^S \geq 0.75$, it is a semi-outdoor space, otherwise, it is outdoor. Thus, the six cases in Figure 2.16, (a), (b), (c), and (d) are classified as O-spaces, whereas (e) and (f) are sO-spaces. It should be noted that C^S of (c) can be more than 0.75 when its length larger than three times of its width. If so, space is also classified as an sO-space. This illustration intuitively shows that "urban canyons" spaces, i.e., small roads/alleys spaces surrounded by high buildings, might end up being sO-spaces or O-spaces.

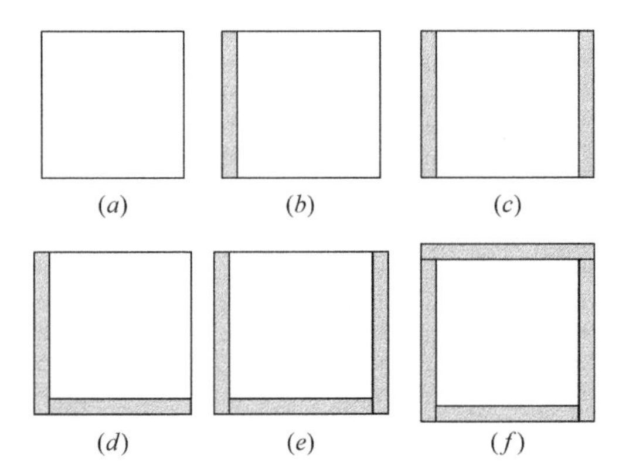

Figure 2.16 Top view of length of surrounding boundaries of square spaces, where slash-filled parts are concrete walls. (a) no wall; (b) one wall; (c) and (d) two walls; (e) three walls; (f) four walls.

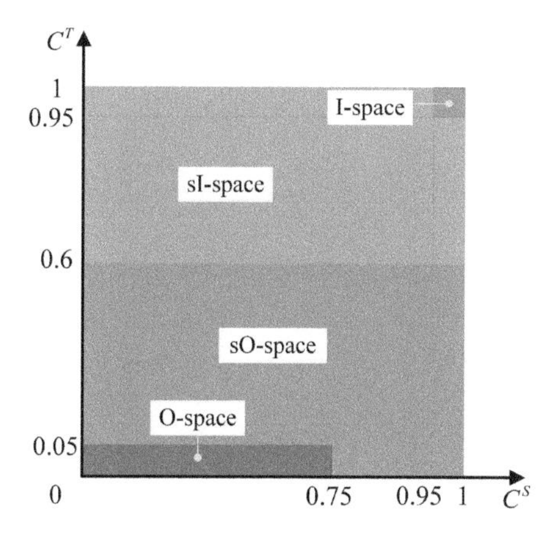

Figure 2.17 Space classification based on the example values of five thresholds.

Example of Quantitative Definitions The five thresholds ($\alpha = 0.05$, $\beta = 0.95$, $\gamma = 0.75$, $\delta = 0.95$, and $\eta = 0.6$) imply that if a space has a top ($C^T \geq 0.6$), it is I-space or sI-space. Then, if $C^T \geq 0.95$ and $C^S \geq 0.95$, the space is I-space, otherwise, it is sI-space. If a space has a top with $C^T < 0.6$, it could be a sO-space or O-space. Further, if its $C^T < 0.05$ and $C^S < 0.75$, it is an O-space, otherwise it is a sO-space, see Equation 2.3.

$$S(C^S, C^T) = \begin{cases} O & \text{if } C^S < 0.75 \text{ and } C^T < 0.05 \\ sO & \text{if } C^S \geq 0.75 \text{ and } C^T < 0.05 \\ & \text{or } 0.05 \leq C^T \leq 0.6 \\ sI & \text{if } C^S \leq 0.95 \text{ and } C^T > 0.6 \\ & \text{or } 0.6 \leq C^T < 0.95 \\ I & \text{if } C^S \geq 0.95 \text{ and } C^T \geq 0.95 \end{cases} \tag{2.3}$$

The effect of these example thresholds on the boundaries between the different spaces is shown in Figure 2.17. The normalization of the coefficients C^T and C^S gives the wrong impression that I-space and O-space are very insignificant and most of the spaces can be classified as sO-space and sI-space. As seen in the examples below, many of the real-world cases have closure close to 1. This means that the orange and green areas in the figure, although showing up as very large, represent a relatively small number of real-world cases.

It should be noted that these values are examples and further research is needed to estimate the effect of closure in diverse situations and environmental conditions, and specify them more accurately.

Classification of Several Real Cases Using the introduced five thresholds, the three spaces in Figure 1.14 can be classified, in which the pergola and eaves are classified as sI-spaces while the roof terrace as sO-space. Figure 2.18 shows more examples of built scenes with a number of built environments: (*a*) Room;

(a) I-space (b) O-space (c) sI-space

(d) sI-space (e) sI-space (f) sI-space

(g) sI-space (h) sO-space (i) sO-space

(j) sO-space (k) sO-space (l) sO-space

Figure 2.18 Some space examples of built scenes.

(*b*) Open-air; (*c*) Gas station; (*d*) The overhanging roof part; (*e*) Overpass; (*f*) Porch; (*g*) Bus stand; (*h*) Courtyard; (*i*) Roof terrace with outside stairs; (*j*) Yard with fence; (*k*) Road with side fence; (*l*) Stadium. Following the same approach as explained above, their coefficients C^T and C^S are estimated. The results of the computations are given Table 2.4.

Table 2.4
Estimated C^T and C^S of the example structures.

Figure	Structure	C^T	C^S	Space Type
Fig 2.18(a)	Room	1.0	1.0	I-space
Fig 2.18(b)	Open air	0.0	0.0	O-space
Fig 2.18(c)	Gas station	1.0	0.0	sI-space
Fig 2.18(d)	Overhanging roof	1.0	0.42	sI-space
Fig 2.18(e)	Overpass	1.0	0.25	sI-space
Fig 2.18(f)	Porch	1.0	0.83	sI-space
Fig 2.18(g)	Bus stand	1.0	0.65	sI-space
Fig 2.18(h)	Courtyard	0.0	0.88	sO-space
Fig 2.18(i)	Roof terrace	0.0	0.8	sO-space
Fig 2.18(j)	Yard	0.0	0.93	sO-space
Fig 2.18(k)	Road with fence	0.05	1.0	sO-space
Fig 2.18(l)	Stadium	0.0	0.8	sO-space

As visible from the examples, many of the cases can easily be classified if their dimensions and closure are known. However, the reality might be not that simple: the dimensions might be not available or the closure not is possible to be estimated. A good example is a stadium. In this research, two spaces are defined: one above the football field and one (or more) under the stadium roof. However, if the entire space (football field and audience space) is considered as one, the stadium space might appear to be sI-space. Other examples are weakly closed spaces for which it is difficult to conclude on a closure. Figure 2.19 shows more cases of weakly top-bounded spaces for which the top has only a decorative (Figure 2.19(a)) or structural reinforcement (Figure 2.19(b)) function. The spaces below them are O-spaces, due to the low C^T that those structures present. In Figure 2.19(c), space corresponds to an sO-space, although the C^S is made up of vegetation fences and houses.

(a) O-space

(b) O-space

(c) sO-space

Figure 2.19 Structures of space have weak characteristics.

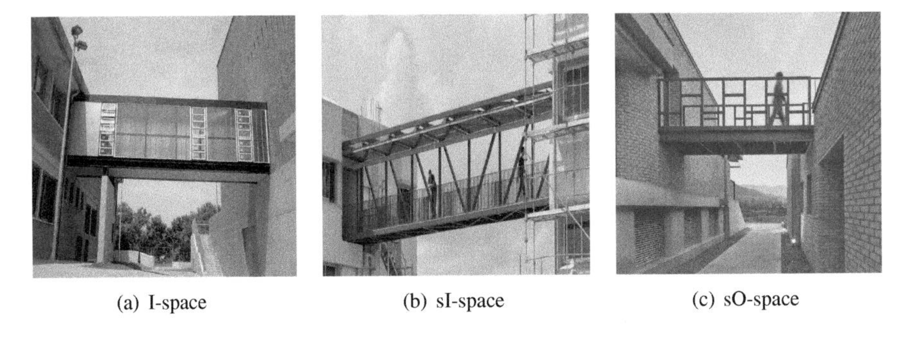

(a) I-space (b) sI-space (c) sO-space

Figure 2.20 Bridges between two buildings.

Furthermore, it cannot be concluded that specific structures can always be associated with the same type of space. For example, Figure 2.20 illustrates the case of a bridge between two buildings. While the first one is an I-space (Figure 2.20(a)), the second one is a sI-space (Figure 2.20(b)), and the third one a sO-space (Figure 2.20(c)). The bridge spaces in (a) and (b) have tops and sides, but the C^T and C^S of the former are nearly 1, while the C^S of the latter is at most 0.9. The bridge space in (c) has no top, but its C^S is more than 0.75. All of the spaces under these bridges correspond to sI-spaces.

2.5 SUMMARY

This chapter presents a summary on space definitions, classifications and representations, followed by an introduction of a generic space definition framework, which is based on the authors' research. Then, the living environments where navigation can happen are classified and defined as four types of spaces: indoor (I-space), outdoor (O-space), semi-indoor (sI-space) and semi-outdoor (sO-space). Following this framework, any space can be classified uniquely into one of the four categories, which provides a basis for integrated and seamless indoor/outdoor navigation. Based on the developed definitions, sI-spaces and sO-spaces can be used in navigation systems, providing the ability to further tune the paths according to the navigating agents' requirements.

The framework controls classifications and definitions of the I-space, O-space, sI-space and sO-space by thresholds of their topClosure (C^T) and sideClosure (C^S), which brings flexibility to the space classification and definition. Thus, it is possible for other space-based models to rely on it and further using the same network extraction approaches across the built environment, thereby providing a seamless navigation solution.

3 Space-based Navigation Models

A navigation model that can represent the spaces and support the path computation is one of the necessary components for navigation. This chapter concentrates on the types and characteristics of space-based navigation models. After introducing the duality theory used for mapping space to a navigation model, existing 2D and 3D approaches for the network model derivation are reviewed. Then, three international standards related to space representation and navigation (namely IndoorGML, IFC/BIM, CityGML) are introduced. Considering it is necessary to consider the relations between navigation model and positioning techniques, the final subsection presents an approach that can derive navigation network for QR code-based indoor navigation, which bridges the derivation approach to positioning techniques.

3.1 NAVIGATION NETWORK

A navigation model is a specific type of data structure, which allows path planning algorithms to be executed [10]. It considers the spaces that are available for navigation and the connectivity between them. Following the two types of representations, B-rep and voxels, two kinds of navigation models can be distinguished as well: network-based navigation models (vector) and grid-based navigation models (raster) [156]. In the first case, the spaces for navigation are further subdivided into voxels and connectivity is established via neighboring navigable voxels. In the case of a network model, the spaces are approximated with nodes and the connectivity is given explicitly by the edge connecting the nodes. This book focuses further on the network-based navigation model. The network models are considered more efficient in terms of processing time, which is essential for large scenes both indoor and outdoor space [157]. As mentioned above, the network-based navigation model is composed of nodes and edges, in which nodes indicate the locations/places (e.g., landmarks or decision points) and edges between them are their spatial interrelationships. Moving from one node to the other is allowed only when there is an edge between them. The network models can be logical and geometric. The logical models provide only connectivity and do not provide geometric location of nodes and the geometric characteristic of the edges. A cost of edges is introduced that indicates the minimizing parameters, most commonly distance or travel time [151]. The nodes can carry information about name, type or function or space. Commonly, a network model has less detailed routing assuming human intelligence will compensate for inaccuracies [3].

Most outdoor navigation models (networks) rely on the available road/street network for vehicles [10, 34]. In contrast, there is a variety of approaches for indoor navigation network derivation, which are based on the geometric, semantic and

DOI: 10.1201/9781003281146-3

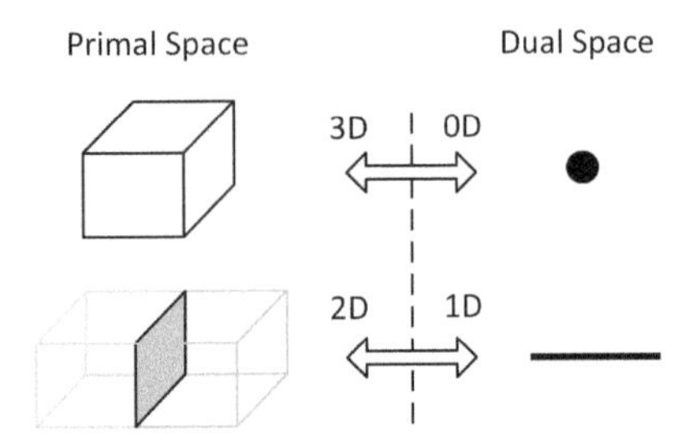

Figure 3.1 Poincaré duality theory used for navigation network derivation.

topological information of indoor (2D/3D) models (e.g., floor plan, BIM model). Approaches for indoor navigation derivation can be summarized into two types: 2D approaches and 3D approaches.

3.1.1 THE POINCARÉ DUALITY THEORY

The Poincaré duality [150] provides a theoretical background for mapping (3D) space to Node-Relation Graph (NRG) [158] representing topological relationships, i.e., it is the theoretical background of simplifying the complex spatial relationships between (3D) objects by a combinatorial topological network model. As seen in Figure 3.1, a k-dimensional object in N-dimensional primal space is mapped to a (N-k)-dimensional object in dual space. 3D solid objects in primal space are mapped to vertices (0D) in dual space. The common 2D face shared by two solid objects is transformed into an edge (1D) linking two vertices in dual space. Thus, edges of the dual graph represent adjacency and connectivity relationships.

The Poincaré Duality has been utilized as the theoretical basis for indoor (3D) space-based navigation, such as IndoorGML [30]. Based on this theory, (3D) spaces (e.g., rooms within a building) are abstracted as nodes, and their adjacency spatial relationships are represented as edges. Note, doors, windows, or hatches between rooms in primal space may also be spaces and therefore might be approximated as nodes. Furthermore, only a subset of the spaces might be used, e.g., only spaces that are on the third and fourth floor can be included in the navigation network. For navigation not all adjacency relationships are utilized, but only those which allow passing from one space to another. The possibility for passing can be estimated via the space semantics, i.e., if two "navigable" spaces are "adjacent", they are connected. The connectivity can be influenced by the accessibility of spaces for specific users. For example, two navigable spaces (room and door) might be adjacent but the door is not accessible (because it is locked) and therefore the user cannot advance through the door.

3.1.2 APPROACHES FOR 2D NAVIGATION NETWORK DERIVATION

2D indoor navigation models represent the indoor spaces and spatial relationships in a simplified way. They carry sufficient information for supporting indoor navigation.

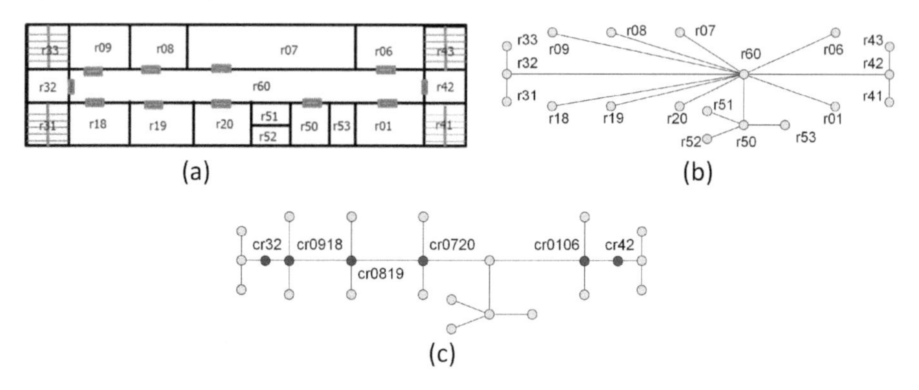

Figure 3.2 A 2D network derived on the basis of Poincaré Duality, MAT and information about doors. The figure is adapted from paper [159]. (a) Floor plan; (b) The metric network derived from connections between spaces; and (c) The corrected geometric network in which the corridor edge is split with respect to the door locations.

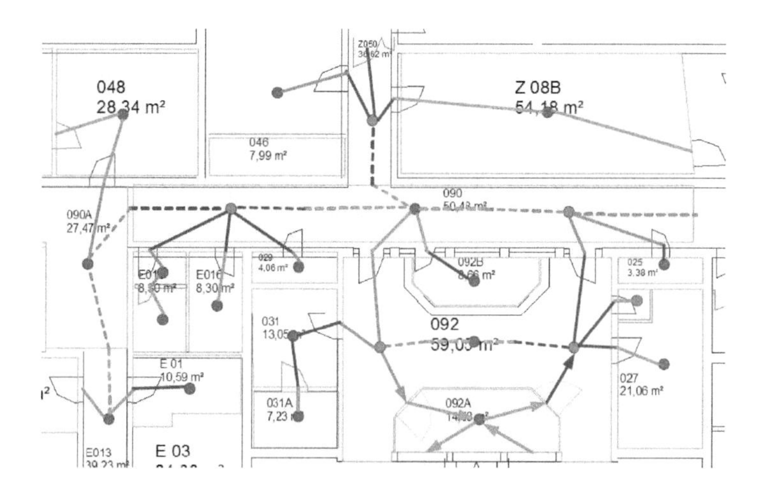

Figure 3.3 The 2D navigation network derived on the basis of room and door spaces [160].

[159] derives 2D network from 2D floor plans on the basis of Poincaré Duality, Media Axis Transform (MAT) and information about doors (Figure 3.2).

Another approach to achieve more accurate paths especially in larger areas to include doors as spaces in the navigation network [160], i.e., both rooms and doors are to be considered as spaces and then abstracted as nodes (Figure 3.3). The long edges that might be created in corridors are symmetrically subdivided into several virtual spaces. [162] derives a 2D navigation network with the similar modification, but more detailed semantics are assigned to the navigation nodes.

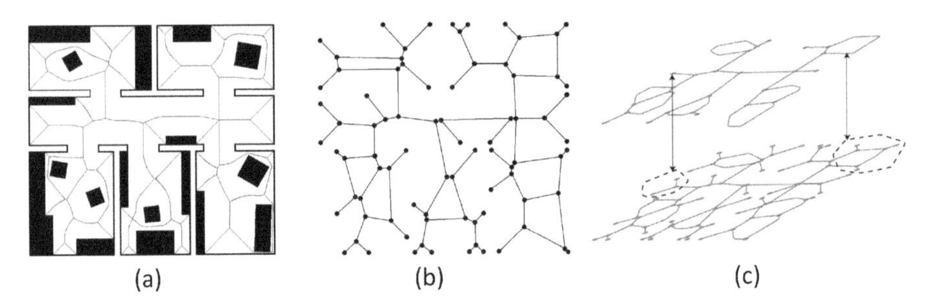

Figure 3.4 The navigation network created on the basis of Voronoi diagram [161]. (a) original plan, (b) network, and (c) simplified network.

The above-mentioned approaches have the an assumption that the indoor spaces are empty (i.e., there is no furniture). However, the same approaches can be used to consider obstacles as well. For example, [161] develops an approach for navigation network derivation based on the generalized Voronoi diagram (Figure 3.4). Space-based approach presented by [163], which considers only doors as spaces to derive a navigation network, computes geometric edges to connect door nodes by estimating both obstacles in the spaces and size of the user. Figure 3.5 illustrates the network model with respect to different user size.

3.1.3 APPROACHES FOR 3D NAVIGATION NETWORK DERIVATION

3D navigation models represent the 3D spaces and spatial relationships between them in horizontal and vertical direction. Deriving 3D navigation models based on 3D spaces has been well-studied in indoor navigation research such as multi-layered space-event model [147], 3D object-based navigation [148], indoor navigation [37, 85], indoor and outdoor combined route planning [32], route computation for emergency response [35].

The authors of [148] present an approach to derive 3D navigation network based on the geometric, semantic and topological information of 3D models. This approach abstracts each separate room, corridor or stair cage as a node based on the Poincaré duality. Then the network is derived by removing weak connections from complete graph, in which the weak connections means the no passage links.

Similar research on deriving networks is based on BIM models [32]. Because this approach generates very rough networks, [37] proposed a method to extract indoor navigation network considering indoor furniture and the navigable spaces. Such a navigation network can support very detailed indoor path planning.

A quite special approach to 3D space is embedding 2D navigation networks into 3D spaces and linking them as 3D navigation networks, which is usually for regular buildings. For instance, [164] generates a network by stacking up 2D navigation networks on each floor, in which the 2D navigation networks are linked by adding connections between the points, where the elevator and staircase are. However, this method is not a true 3D approach [3].

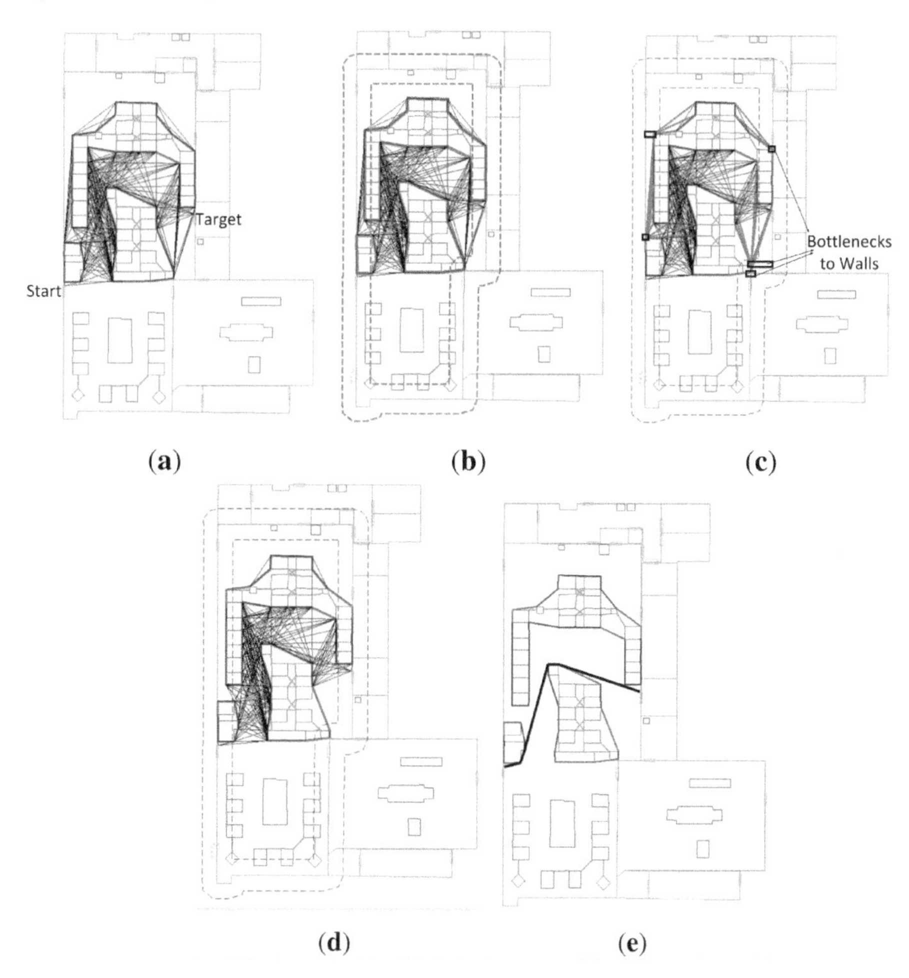

Figure 3.5 Creating a visibility graph (VG) and a room buffer, removing the inaccessible edges and computing a path for users with a size of 0.8 m. (a) Creating a VG; (b) computing a room buffer; (c) finding inaccessible edges in the bottlenecks; (d) removing the inaccessible edges; (e) the computed path. This figure is borrowed from paper [163].

3.2 INTERNATIONAL STANDARDS RELATED TO NAVIGATION

3.2.1 INDOORGML

The standard IndoorGML [30] is an international standard of Open Geospatial Consortium (OGC). It is an open data model and has XML-based schema, but for indoor needed to support navigation applications. IndoorGML aims to provide a common framework for representation and exchange of indoor spatial information based on 3D spaces and network/graph. This standard is the most well-known standard for indoor navigation based on 3D space-based navigation model.

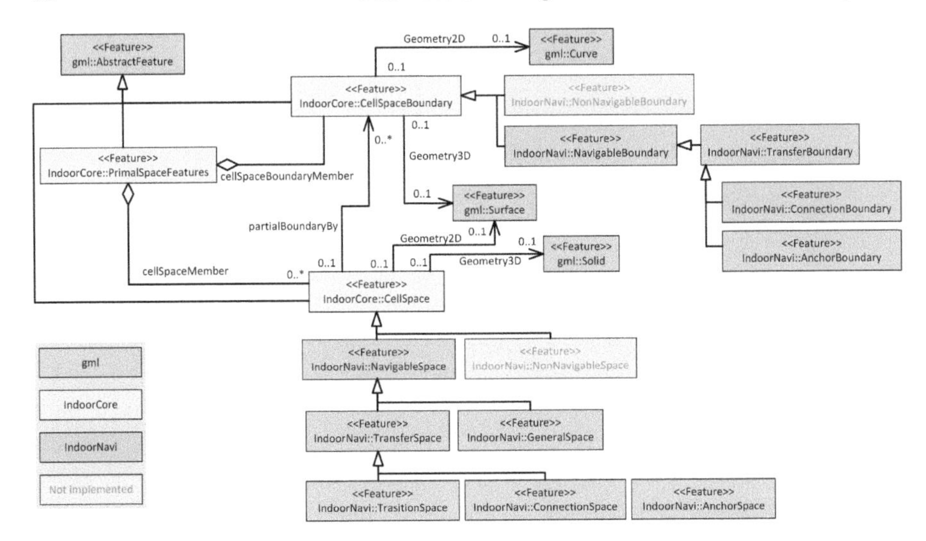

Figure 3.6 Part of IndoorGML 1.0.3 core and navigation module UML diagram (OGC, IndoorGML [30]).

The navigation module of this standard provides semantic information for indoor space to support indoor navigation applications. As seen in Figure 3.6, the information is organized in two modules: core and navigation. The core module provide the basis classes for the space and network model. The two major classes in the core module defining spaces are *CellSpace* and *CellSpaceBoundary*. A *CellSpace* is a semantic class corresponding to one space object in the Euclidean primal space of one layer. A *CellSpaceBoundary* is used to semantically describe the boundaries of each space object. Space features are further classified in the Navigation module into two groups: *NavigableSpace* and *NavigableSpaceBoundary*. *NavigableSpace* represents all indoor spaces (e.g., rooms, corridors, windows, stairs) that can be used by a navigation application. *NavigableBoundary* represents all geometries that border the spaces but can be passed, such a the polygon between a room and a door space.

IndoorGML defined the space as a space cell that is bounded by physical or virtual components, which is similar to the common understanding that indoor space is inside buildings (such as a house, or a commercial shopping center), where people usually are able to move. Figure 2.10 illustrates an indoor scene with two rooms and two doors. Then all the building components(walls, doors) and indoor spaces themselves are represented, in which $B1$ to $B8$ are walls, $D1$ and $D2$ are doors, while $CellR1$ and $CellR2$ are two indoor spaces.

Presently, IndoorGML covers geometric and semantic properties relevant for indoor navigation in an indoor space. These spaces may differ from the spaces described by other standards such as CityGML and IFC. In this respect, IndoorGML is a complementary standard to CityGML and IFC to support location based services for indoor navigation. It should be noted that although some modifications have been proposed for the IndoorGML navigation model (e.g., [165]), this research concentrates on the current version 1.0.3 of the standard.

3.2.2 INDUSTRY FOUNDATION CLASSES (IFC)

The most well-known international standard that represents spaces as 3D objects is probably the Industry Foundation Classes (IFC) standard, developed by the International Alliance for Interoperability (IAI) [166]. In this standard, space is an area or a volume that is bounded by physically or imaginary elements. Spaces are intended to describe certain functions of the building. Commonly, a space in the IFC standard is associated with a building/building storey or a room, which indicates that space is interior. However, it can also be associated with a construction site, case in which that space is exterior. The space definition in Building Information Modelling (BIM) serves therefore for a large number of purposes including calculations of energy use, acoustic analysis, navigation, orientation within the building, egress simulations. Property measurement and valuation-related standards define various types of interior areas and volumes [167]. These different types of area and volume definitions show various different interpretations of interior spaces in the domain of property and facility management.

The IFC standard is able to accommodate both geometric information and rich semantic information of building components to support indoor navigation. Thus, IFC building models have been used for indoor navigation [84, 168–170]. [168] achieved the shortest path planning for 3D indoor spaces based on IFC, in which they extracted both geometric and semantic information of building components that defined within the IFC file, including *IfcSpace*, *IfcWall*, *IfcDoor*, etc.

3.2.3 CITYGML

City Geography Markup Language (CityGML) [171] is an OGC standard for an open data model and XML-based format for the storage and exchange of virtual 3D city models. It provides a generic semantic, attributes and relations of a 3D city surface model. This is especially important for cost-effective sustainable maintenance of 3D city models, allowing reuse of data in different application fields. In this standard, a large number of indoor and outdoor physical entities are given digital representations. Furthermore, One of the targeted application areas of CityGML is vehicle and pedestrian navigation.

In the current version 2.0 of CityGML, spaces are not explicitly mentioned, but notations as rooms, doors, windows, being represented as volumes indicate that indoor spaces are critical. The road areas consist of traffic areas and auxiliary traffic areas (Figure 3.7). The former includes pedestrian pavements and vehicle lanes, while the latter consists of grass areas and tree areas. In the CityGML data model, several classes available in that standard, make it interesting to work with for the purpose of detecting the semi-bounded spaces with upper boundaries. Building elements like balconies, chimneys, dormers or outer stairs are represented by the *BuildingInstallation* class. The shell based representation provided by the standard gives direct access to interesting components such as *RoofSurface*, *GroundSurface*, *OuterCeilingSurface* and *OuterFloorSurface* (Figure 3.8). Those latter elements are allowed to be encapsulated or referenced by the *boundedBy* property of *BuildingInstallation*,

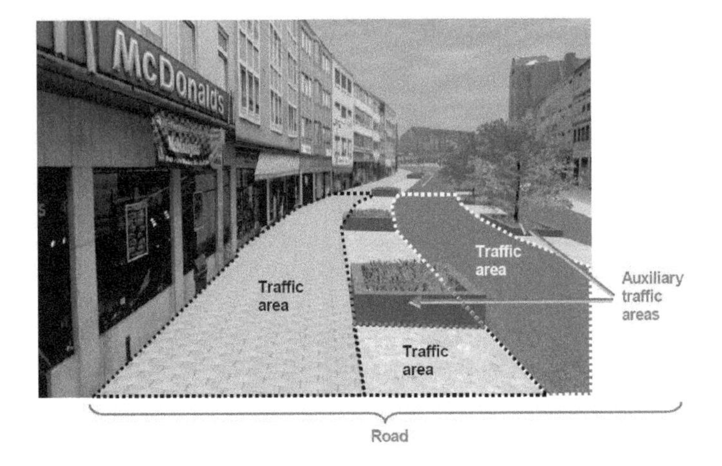

Figure 3.7 Road areas consist of traffic areas and auxiliary traffic areas [171].

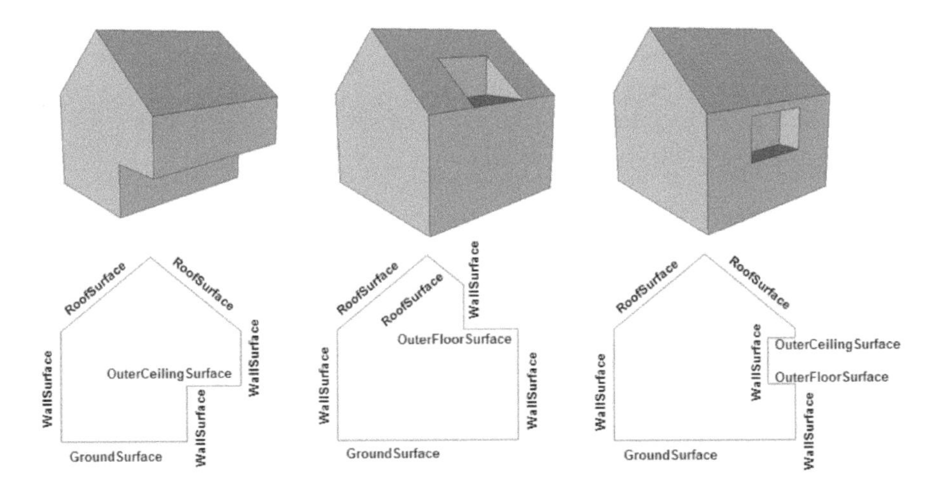

Figure 3.8 Buildings components in CityGML shell model [171].

along with *WallSurface* and *ClosureSurface* elements. Other elements of interest, such as *FloorSurface*, *CeilingSurface*, *InteriorWallSurface*, and *ClosureSurface* are more relevant to the indoor spaces. For this reason, they are allowed to be encapsulated or referenced by the *boundedBy* property of the *Room* class.

Despite CityGML version 3 has been in an very advanced stage, this research concentrates on version 2.0 of the standard. Spaces as a generic class in CityGML 3.0 are almost finalized [172]. A road is represented to have a driving lane and two sidewalks (Figure 3.9). Then, the free spaces above the driving lane and the sidewalks are traffic spaces (in blue); whereas the ground surfaces (in green) are utilized as the lower boundary of the traffic spaces. Moreover, the height of the traffic spaces

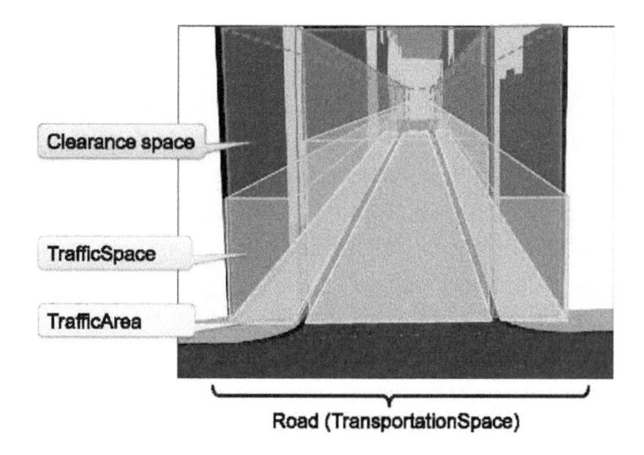

Figure 3.9 Representation of a road by 3D spaces in CityGML 3.0 [172].

and sidewalks spaces are set as 4.5 m and 2.5 m respectively based on the standard heights of roads in Germany. In addition, it also represents the clearance spaces in red.

3.3 NAVIGATION NETWORK DERIVATION FOR QR CODE-BASED INDOOR NAVIGATION

The above-mentioned approaches for navigation network derivation as well as the three related to spaces standards are completely independent from positioning techniques. In practice, the relations between navigation network and positioning techniques should be taken into consideration, because if a navigation network is not appropriate to the positioning technique, some ambiguous issues may appear. Among the indoor positioning, QR (Quick Response) code is one of the low-cost, easily deployable, flexible, and efficient approach, which has been often used for indoor positioning and navigation purpose. Therefore, taking QR codes as an example, this section illustrates an approach that considers the specifics of the indoor positioning approach.

3.3.1 QR CODE-BASED INDOOR NAVIGATION

QR codes are matrix codes, which are similar to two-dimensional bar codes, but allowing for more flexible encoding a large spectrum of information, quick decoding information. They are readability from any direction from 360 degrees [173]. More importantly, QR codes can be read by hand-held devices with cameras, e.g., smartphones. Compared to other indoor positioning techniques, QR codes have many advantages, e.g., (a) high positioning accuracy as QR codes directly carry the precise location information (coordinates); (b) anti-interference, because they are independent of radio signal strength information (RSSI); (c) strong flexibility as quantities and the locations of QR codes can be easily adjusted; (d) low cost for installation and maintenance, since production cost of QR codes is very low, i.e., only for printing the

QR codes on appropriate material and installation; and (e) no special infrastructures are needed, because accessing location of QR codes is readily available on smartphones. Depending on the material they are printed on, QR codes can be vulnerable to damage, nevertheless, the cost of replacing damaged QR codes is very low. Therefore, we deem that QR code is a very promising positioning method that could be popularized and used in a large scale indoor positioning and navigation.

In the past decades, research on QR-code based indoor navigation has mainly focused on designing the architecture of the navigation system based on the QR codes [174, 175], encoding/decoding of the QR codes [176–178] and combining QR codes with other techniques for navigation [179, 180]. [173] developed an indoor navigation technique for finding the optimal route for visually impaired using QR codes. In the implementation, they placed horizontal strips of QR across the full width of the path (corridor) for error-free scanning of QR codes. In the research of [176], whole floor maps are encoded as QR codes, and the generated QR codes are placed at the entry places and vital places in the building. Users can get recovered floor maps by scanning QR codes. [181] placed QR codes on the significant or key places/locations. [178] presented a geo-coding framework for indoor navigation and designed a QR code-based indoor navigation system, in which the indoor spaces are decomposed into internal Point of Interest (POI), where a POI is a room or a specific functional area. Namely, POIs are the key objects, and location information at each POI is recorded in form of the QR codes. [180] indicated that QR codes are placed along the pathway and authors of [174] pointed out that the QR codes should be positioned on the characteristic indoor places to allow users to determine their positions on the map.

The process of path planning by leveraging QR codes is as follows: (i) in an application installed on the user phone, the user provides the target destination, (ii) the user walks to the closest visible QR code and scan it, (iii) the application decodes the location and searchers for the closest node of the navigation network, which is considered as a start point and (iv) the application computes a path between start and target point. This section presents a navigation network derivation approach to address the issue by integrating the QR code locations as nodes in navigation networks.

Commonly, the indoor navigation network is derived independently from the QR-code localization. For instance, no thorough research has been completed to investigate the relation between navigation networks and locations of QR codes. Such practice may cause ambiguities when deciding the closest node from the network that should be used for path computation. Specifically, QR codes are generally placed according to preferences or certain specifications and navigation networks are derived without considering QR code locations, In other words, depending on the configuration of the network and locations of the installed QR codes, the application might be not able either to find the closed node of the network, or could find a wrong node that can falsely place the user even in a wrong room. A common method to search for the closest nodes is to set buffer areas for QR codes. Figure 3.10 illustrates a case of a corridor and automatically created navigation network. The buffer created at the location of QR code at POI2 is not able to find a node of the network. In contrast, the buffer created around POI3 finds two nodes of the network. Similarly, based on

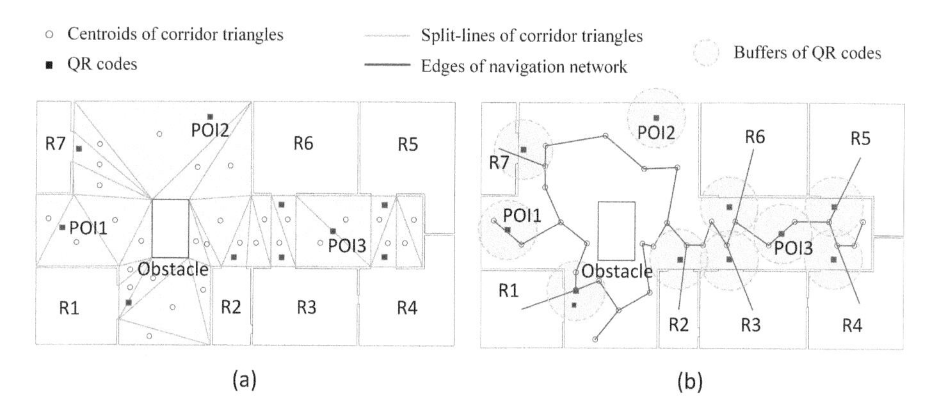

Figure 3.10 Navigation network derivation based on CDT. (a) Space subdivision of corridor based on CDT; (b) navigation network derivation based on Poincaré duality [150].

the buffer of the QR code in front of the door of R7, it is confusing to determine it is in R7 or not. To avoid such kinds of ambiguity, we argue that the relations between navigation network and QR codes should be taken into account when deriving a network, such as include the locations of the QR codes as a part of it.

Inspired by the space-based navigation network generation methods discussed above, we propose to associate QR codes with spaces. To achieve this, we use the 2D space-subdivision approach Voronoi tessellation (also called Voronoi diagram) [182]. The subdivision principle relies on a set of seed points (also called generators or sites). With the help of these points, a plane can be subdivided into regions, so that each region of a seed point, which is composed of set of points at the edges and vertices of the region, is at least as close to the seed point of the region as to any other neighboring seed point. Once taking the locations of QR codes as additional generators to subdivide indoor spaces, a region around a QR code will be created, which will contain only one QR code. This will ensure the QR code can be considered in the navigation network. In other words, this paper employs Voronoi tessellation as the fundamental mathematical approach to subdivide the indoor and further build navigation networks.

3.3.2 INDOOR SCENE CLASSIFICATION

Indoor spaces are commonly perceived as the places that are bordered by physical unmovable elements (e.g., walls, ceilings, floors). The spaces are semantically identified as room, corridor, stair, lift, toilet, which largely corresponds to the functions or usages of spaces [3, 30, 183]. In indoor pedestrian navigation, humans are more accustomed to using semantic information rather than coordinates. An example of a navigation path request can be that: someone needs a path from Entrance A to Room 4036, rather than a path from $P(x_0, y_0, z_0)$ to $P(x_1, y_1, z_1)$. Furthermore, pedestrians only can visit the spaces that are free of obstacles and accessible [12]. Therefore, for

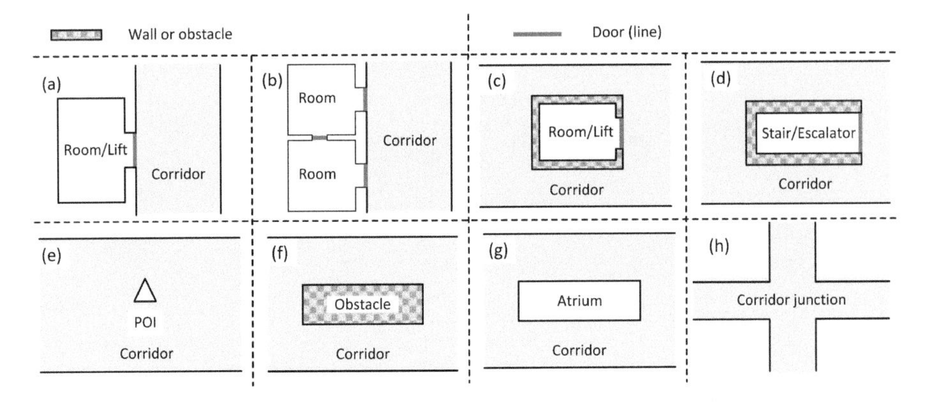

Figure 3.11 Classified and simplified indoor scene.

the purpose of illustrating QR-code based navigation, we introduce semantics for the indoor free spaces.

On the basis of indoor semantics, we classify the indoor scene into five categories: room-like space, corridor, atrium, indoor POIs, and obstacle. As seen in the Figure 3.11, the room-like spaces connect with the corridor or other rooms by door(s) (which are simplified as lines), the atrium, indoor POIs, and obstacle are laying inside of the corridor. In this case we do not associate the doors with spaces. The definitions of the five categories are the following:

+ Room-like space. Indoor spaces, in which people spend a longer period of time to work, shop, watch, dine etc. Such spaces are connected to corridor or other room-like spaces by doors or openings, or vertical connectors such as lifts, stairs, escalators, etc.
+ Corridor. Corridors are the passageways in indoor scenes, which are used to move from one room to another or to different storeys. The corridors are normally used for a very short period of time although if benches are provided people may use them to rest or wait for friends and family members.
+ Atrium. Atriums are defined as large open spaces within buildings (e.g., the open space in a shopping mall in Figure 3.12). They are often several stories high and having glazed roofs or large windows, and located immediately beyond the main entrance doors (in the lobby).
+ Indoor POIs. An indoor POI is a location in indoor space where information regarding a particular place, service, facility, or event is available, in contrast to traditional POIs located in outdoor environments [184]. Indoor POIs are provided to support various indoor applications and services, including route navigation, facility management, and evacuation simulation. In this book POIs are merely the facilities that pedestrians may be interested in, such as ATM, bench, fixed-position table. Moreover, all indoor POIs are abstracted as points regardless the original shapes of them.

Figure 3.12 Atriums in a shopping mall.

+ Obstacle. Obstacles are spaces occupied by physical objects that pedestrians are unable to go through, such as wall, pillar, and hand railings and furniture. In this QR-based example for network derivation, obstacles will not be considered.

3.3.3 SPACE SUBDIVISION AND NAVIGATION NETWORK DERIVATION

QR code placement in the physical environment is critical, because they will be used as the seeds in Voronoi tessellation for corridor space subdivision. They could be placed on the significant or key places/locations [181], along the pathway [180], or on the floor of the corridor [173]. For this work, we have decided to place the QR codes on the ground, because we were advised this has been a common practice for many shopping malls. In particular, the following two practical recommendations were followed: (i) QR codes should be placed close to the places, where pedestrians walk, and (ii) they should be easier for users to find and scan them, especially within large open spaces. For room-like spaces, they are placed on the floor at a certain distance at front of the doors but outside of the rooms, on the side of the corridor; If the QR-codes are also POIs, they are placed on the exact locations of POIs (Figure 3.13). Using QR code locations as seeds in the Voronoi tessellation, results into 2D spaces, which can be directly used to derive a navigation network. After removing the atrium/obstacle areas, we can obtain the subdivided corridor, in which each QR code corresponds a subspace (Figure 3.13(a)). Afterwards, a navigation network is automatically derived with the support of Poincaré duality (Figure 3.13(b)). It should be emphasized that the nodes extraction in this paper has two cases: (i) if a subspace contains a QR code, the location of the QR code will be used as the node to represent this subspace, such as corridor subspace; (ii) if a subspace does not contain a QR code, the centroid of the space will be utilized as the node to represent this space, such as rooms.

We believe the network shown in Figure 3.13(b) is insufficient for pedestrian navigation, because some edges cross corners of walls, obstacles/atriums, see the edges circled by dotted ellipses in that figure. Because of these issues, we may get impractical paths. For instance, from R1 to the POI1, an impractical path for pedestrians will be like that: turn left after leaving R1, go through the walls of R1, then toward the destination.

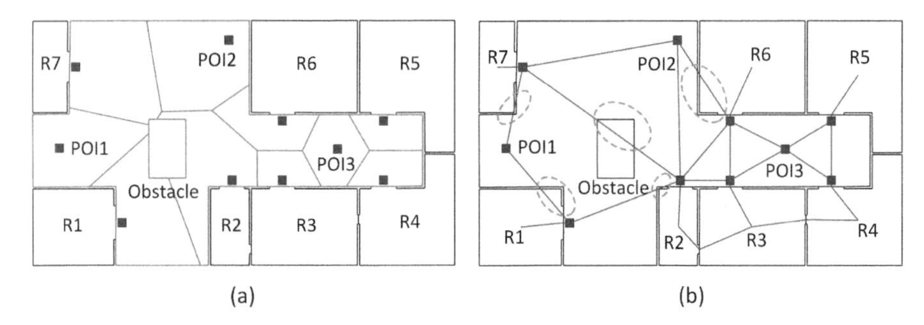

Figure 3.13 Space subdivision and navigation network derivation, in which the black squares are QR codes, green lines are split-lines, blue lines are the navigation edges, and red dashed circles mark positions of the edges crossing walls. (a) QR codes and subdivided corridor; (b) the derived navigation networks based on Poincaré duality.

3.3.4 DUMMY NODES AND EXTENDED NAVIGATION NETWORK

As shown in Figure 3.13(b), the crossing wall issues mainly happen at the corners of walls, obstacles/atriums. Such crossing may result in incorrect guidance and needs to be avoided. We propose a practical approach, which is introducing additional nodes, which do not correspond to spaces, but ensure that the network edges do not cross obstacles. We call such nodes "dummy". The dummy nodes are located at the concave and convex corners. They play similar role as the QR codes in the space subdivision and navigation network derivation, but no physical markers are placed at their locations.

Dummy nodes are automatically computed based on the corners of atrium/obstacle and corridors. However, it does not mean they can be randomly chosen at the corners, since their locations directly affect the navigation network. It is also not a good idea to directly employ the corners as dummy nodes because users usually keep a distance from them during the navigation. Therefore, we put forward two principles for computing dummy nodes: (i) they shall keep a certain distance to the corners [47]; (ii) they shall take into account the two directions of the corners.

Figure 3.14 illustrates the process of computing dummy nodes locations. Utilizing the corner A as an example, Figure 3.15 shows the whole process of computing the locations of dummy nodes (A_i). The computation starts from the two edges that form the corner A. To facilitate calculation, the distance between a corner (A) and auxiliary points (M and N) is set as d. For the edge formed by B and A, to compute the point (M) on the its extension, the slope of this edge will be used, in which $x_A - x_B \neq 0$ is considered as a condition to ensure the validity of the slope calculation. Once this condition is true, the M can be computed by Formula 1, otherwise, Formula 2. There are two points can be obtained for each computation, but only the one is on the extension of the edge will be used for the subsequent calculation. Similarly, another auxiliary point N can be computed based on the edge formed by F and A (Formula 3 and 4). Then, auxiliary point I can be easily computed based on M and

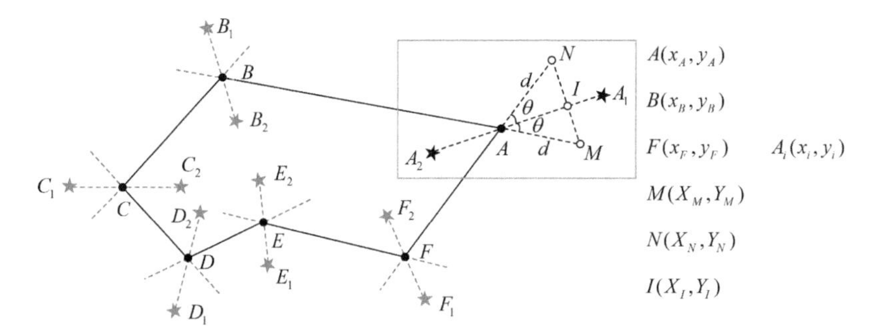

Figure 3.14 Determination of dummy node locations of a space, which can be a corridor or atrium/obstacle. Black solid lines represent walls or boundaries of the space. The space has six vertices with known coordinates, including *A*, *B*, *C*, *D*, *E*, and *F*. A_i are the dummy nodes of corner *A*. *M*, *N* and *I* are auxiliary points for calculation. The stars are potential dummy node locations; dotted lines are auxiliary lines.

N (Formula 5). Finally, using the *A* and *I*, the dummy nodes A_i can be computed, in which we also consider the condition that $x_A - x_I \neq 0$ to make sure the slope computation of the line segment formed by *A* and *I* is valid. Hereafter, if this condition is true, the dummy node is computed based on Formula 6, otherwise, the Formula 7.

By repeating the whole processes illustrated in Figure 3.15, all the corners of the space in Figure 3.14 can be computed, such as dummy nodes for *B* are B_1 and B_2. Only one of the two nodes will eventually be selected, and the selection criteria is if inside or outside nodes are needed. For instance, if this space is an atrium/obstacle, all the outside nodes should be selected, i.e., the dummy nodes are A_1, B_1, C_1, D_1, E_1, and F_1. If this space is a corridor, all the inside nodes are needed (i.e., A_2, B_2, C_2, D_2, E_2, and F_2).

The QR codes and dummy nodes of different indoor scenes are illustrated in Figure 3.16. Utilizing the locations of QR codes and dummy nodes as seeds, the corridor in the example is subdivided by using the Voronoi tessellation. For each QR code or dummy node, there is only one corresponding region. After removing the atrium/obstacle areas by using region difference, we can get the final subdivided corridor (Figure 3.17(a)). Then, using Poincaré duality, an adapted navigation network can be automatically derived (Figure 3.17(b)). The adapted navigation that includes QR codes and dummy nodes to avoid corners, provides simple yet efficient solution to the above-mentioned issues.

3.4 SUMMARY

As discussed in this chapter, network-based navigation model is more commonly used for pedestrian navigation. There are many 2D and 3D approaches for deriving such navigation models. They can be even combined to derive the desired network. The space-based approach is a powerful and flexible concept that allows to consider

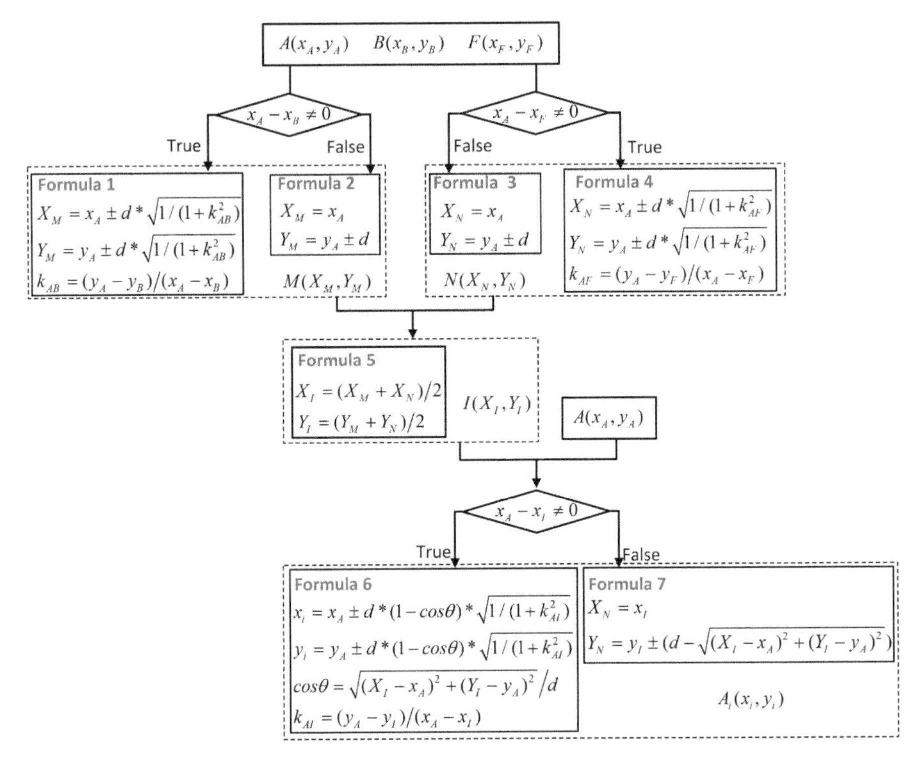

Figure 3.15 Determination of dummy node locations (A_i) based on the corner (A) formed by edges AB and AF.

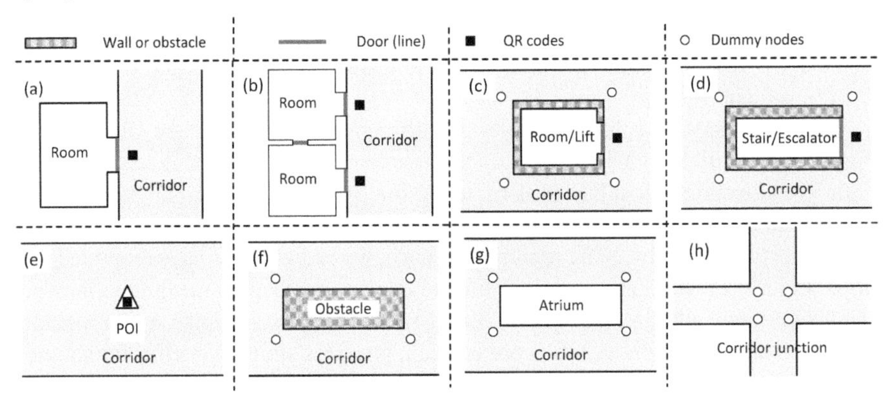

Figure 3.16 QR codes and dummy nodes of all the classified indoor scenes.

not only the building structure but also the purpose of the navigation, and the dimensions of the user.

There are three international standards, i.e., IndoorGML, IFC and CityGML, which represent 3D geometry and are suitable to derive navigation networks. IFC

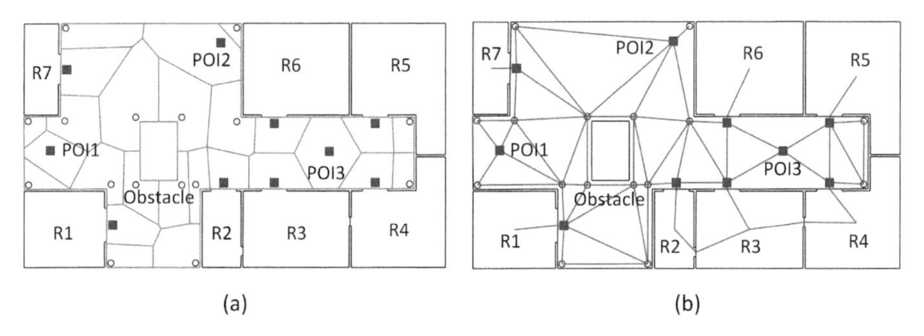

Figure 3.17 Extended space subdivision and navigation network. The black squares indicate locations of QR codes, hollow circles are locations of dummy nodes, green lines are split-lines, and blue lines are the navigation edges.

and CityGML are more general standards to represent 3D indoor environments. IndoorGML is dedicated to space-based navigation. Although the focus is on indoor, IndoorGML concepts of CellSpace, CellSpaceBoundary, Navigable and Non-navigable Spaces and Boundaries has the potential to reflect all possible spaces (indoor, outdoor and semi-bounded) and their boundaries to support seamless indoor/outdoor navigation.

Many applications compute the navigation model independently form the indoor localization system. However, depending on the indoor position technology, the relations between navigation model and localization approach may be important to take into consideration. This chapter illustrated the criticality of such link with a QR code-based localization. Compatible approaches are recommended to be investigated for other localization technology. For example it might be important to estimate the coverage and distribution of tag/receivers for Ultra Wide Band (UWB), Wifi, RFID, iBeacon, etc. systems.

4 Unified Space-based Navigation Model

From a technical perspective, a 3D navigation model is one of the critical components that should be available to perform successful navigation. A common approach to build a unified navigation model to support seamless path computation is linking indoor navigation networks to outdoor road/street-based networks. Because of different sources of indoor and outdoor navigation networks, this approach fails to build up true seamless navigation models. As illustrated in previous chapters for indoor environments, spaces are powerful concept that can be used to provide seamless indoor/outdoor navigation model.

This chapter provides a unified 3D space-based navigation model (U3DSNM). The chapter presents the conceptual model (spatial schema) organized in UML class diagram. The second subsection illustrates the corresponding technical model by mapping its conceptual classes to python classes. The unified navigation model ensures all types of spaces for navigation (indoor, semi-indoor, semi-outdoor, and outdoor) have the same representation, management methods, and network derivation approach, thereby building up unified navigation networks to support seamless navigation paths planning. The model can be linked to the international standards (data models) that are also based on spaces, such as IndoorGML and the on-going version of CityGML 3.0.

4.1 REQUIREMENTS TO A UNIFIED SPACE-BASED NAVIGATION MODEL

Building upon investigations on the indoor navigation requirements [3, 9, 61, 84, 86, 185], seamless navigation and spatial model [10, 13, 15, 186], indoor/outdoor navigation service [34, 80, 187], and considering the generic concepts presented above, we attempt to list five navigation use cases, example scenarios, and required information (Table 4.1). The use cases specify four major requirements for a unified space-based navigation model as follows:

1. Spaces and their boundaries in indoor and outdoor shall be represented in a uniform way, such as *Space* and *Boundary*.
2. Spaces shall be geometrically described and classified according to semantic properties, which define connectivity between spaces. The boundaries shall carry information about their roles (Top, Side, or Bottom), their closure, and their physical consistency because such information can indicate or help agents to determine whether they can employ spaces for some specific purposes. For instance, using tops for escaping rain or strong sun. Similarly, utilizing sides for escaping from strong wind.

DOI: 10.1201/9781003281146-4

Table 4.1
Four use cases to abstract the requirements of seamless navigation.

	Use Case	Example Scenario	Required Information
1	Provide a seamless navigation path for users.	A user wants to find a path from an outdoor space to a specific indoor space.	All indoor and outdoor spaces; a unified seamless navigation network.
2	Provide a most/least-top-bounded navigation path for users.	A user wants to find/avoid structures from the top direction that may shelter the sunshine or rain during the navigation.	Top closure of spaces; type of space; roles of boundaries.
3	Provide a most-side-bounded navigation path for users.	A user wants to structure from surrounding directions during the navigation to escape from strong wind.	Side closure of spaces; type of space; roles of boundaries.
4	Provide a navigation path that uses the size of users as constraints.	A user needs a navigation path that can consider the size of accompanying equipment.	The size of users; size of all virtual boundaries; 3D geometry of spaces.
5	Provide a navigation path to a certain obstacle.	A user wants to find a printer for printing.	The obstacle spaces; semantic of spaces.

3. Navigation networks shall be derived uniquely and automatically from the above-defined spaces utilizing Poincaré duality [150]. Nodes in the network indicate spaces and edges between them annotate the connectivity of spaces.
4. Obstacles spaces shall be represented distinguishably as non-navigable spaces. They shall be semantically categorized as static (e.g., pillar, tree), semi-mobile (e.g., small table, car), and mobile (e.g., fire) [37].

4.2 CONCEPTUAL MODEL OF UNIFIED 3D SPACE-BASED NAVIGATION MODEL (U3DSNM)

As stated in the beginning of this chapter, a possible solution of building up a unified navigation model (network) for seamless navigation is to develop it based on (3D) spaces used for navigation. We built the model around the core notations Primary

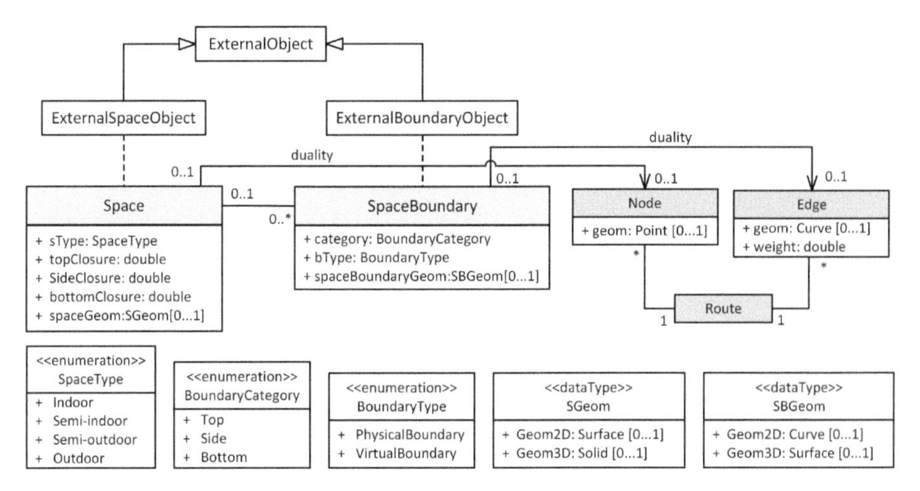

Figure 4.1 UML diagram of the U3DSNM.

and Dual Space of IndoorGML as discussed in Chapter 3. The concepts of space, space boundary, and the related navigation network are represented in the Unified Modeling Language (UML) diagram of the U3DSNM by introducing five classes *Space*, *SpaceBoundary*, *Node*, *Edge*, and *Route* as shown in Figure 4.1.

The four types of space and three closures are organized as attributes of *Space*, namely, *sType*, *topClosure*, *sideClosure*, *bottomClosure*. The attribute *sType* distinguishes spaces for navigation into four types, in which its data type is set as *SpaceType* with enumerated values: Indoor, Semi-indoor, Semi-outdoor and Outdoor. Although it seems that the three types of closures are boundary characteristics, they are stored as attributes for *Space*, because they are globally characterizing the average proportion of solid (material) around a *Space*. Based on the definitions of closures, the data type of the them is a decimal between 0 and 1. Moreover, the geometry of a space is represented as an attribute named *spaceGeom* with a data type of *SGeom*, which means a space is represented as a surface in 2D while a solid in 3D geometry.

Similarly, three roles of boundaries with two types are organized as two attributes of *SpaceBoundary*, in which the roles of boundaries are discriminated by an enumerated attribute *category*. As the definition of the attribute implies, this attribute indicates roles of boundaries, which thus having enumerated values: Top, Side and Bottom. Another attribute (*bType*) is introduced to distinguish the physical characteristics of a boundary, which is enumerated as PhysicalBoundary and VirtualBoundary. The geometry of a space boundary is represented as an attribute named *spaceBoundaryGeom* with a data type of *SBGeom*, in which the 2D geometry of a boundary is a curve while that of 3D is a surface.

The link between the *Space* and *SpaceBoundary* indicates that a space object can have several boundaries but a boundary only belongs to one space or shared by two spaces. For instance, a room can have several walls but a wall belongs to one room or is shared by two rooms.

Table 4.2
Example of sI-space table in the computer memory.

topClosure	sideClosure	bottomClosure	containBoundary
1	0.49	1	[b0, b4, ...]
...	...	...	...

The Duality Space, which describes the navigation network, is represented by three classes, namely *Node*, *Edge*, and *Route*. Following teh duality principle, a space is linked to a node and a space boundary to an edge. It should be noted that only the virtual boundaries that are shared by two spaces are mapped as edges, because agents cannot go through physical boundaries in navigation. Combining nodes and edges, navigation network can be established, which further be utilized for route computation.

4.3 TECHNICAL MODEL: PYTHON CLASSES

The UML model of U3DSNM is mapped into tables, which are maintained in the memory. The same memory data structure can be implemented in Relational Database Management Systems (RDBMS). The *Space* in the UML model is mapped as four tables to represent four types of spaces. Table 4.2 illustrates the data structures of the semi-indoor spaces in the computer memory. The closures are transferred as three columns with the same names of the attributes from *Space*. Another column named *containBoundary* is added to store the IDs (e.g., "b0", "b3") of the boundaries that form a sI-space, further representing the geometry of the space. Thus, each sI-space is represented as a quintuple: sI-space = {topClosure, sideClosure, bottomClosure, containBoundary}. All closures of I-spaces are directly set as 1 because all of them come from BIM model. Although the IDs of space boundaries are not shown in the UML, we consider each object will default have an ID.

The *SpaceBoundary* in the UML is also mapped as a table (Table 4.3), which means each boundary is represented as a quintuple: boundary = {category, bType, SBGeom, fromSpace}. All the three attributes from the UML are utilized as three columns with the same names. Besides, an extra column named *fromSpace* is added to indicate this boundary comes from which space, in which the IDs of spaces (e.g., "sI1", "sI2") are stored.

4.4 MAP TO INDOORGML AND CITYGML

The U3DSNM can easily be mapped to the international standards (data models) that are also based on spaces, such as the working versions of IndoorGML 2.0 [30] and CityGML 3.0 [172]. This model enriches the two standards with more semantics, which makes them more powerful and comprehensive, especially for navigation.

Table 4.3
Example of *PhysicalBoundary* table in the computer memory.

category	bType	SBGeom	fromSpace
Top	Physical	$[p_0,p_1,...,p_0]$	sI1
Side	Virtual	$[p_{13},p_7,...,p_{13}]$	sI5
Bottom	Physical	$[p_{22},p_8,...,p_{22}]$	sI2
...	...	...	...

From another point of view, U3DSNM can use the data based on the two standards as inputs.

Figure 4.2 shows the UML diagram that integrates the classes resulting from the U3DSNM to existing classes of IndoorGML. The *Space* and *SpaceBoundary* are respectively mapped to *NavigableSpace* and *CellSpaceBoundary* to enrich the contents of the two classes in IndoorGML, because the I-spaces in U3DSNM are exactly the indoor navigable spaces in IndoorGML, so does the SpaceBoundary and

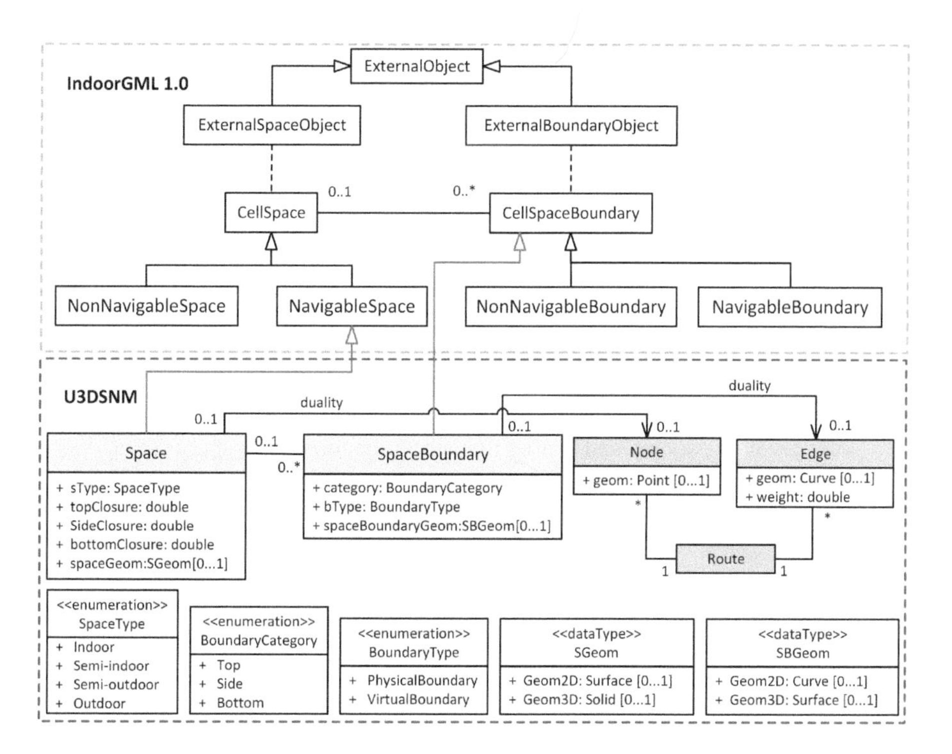

Figure 4.2 Map U3DSNM to IndoorGML 1.0.

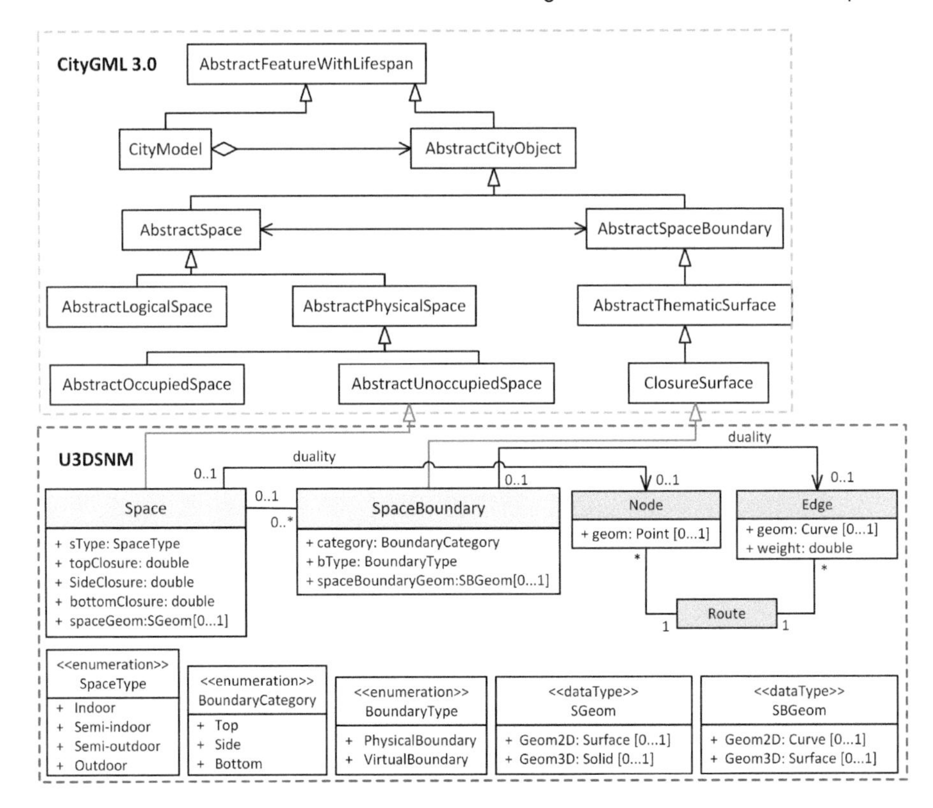

Figure 4.3 Map U3DSNM to CityGML 3.0.

CellSpaceBoundary. In IndoorGML, a *CellSpace* is a semantic class corresponding to one indoor space object in the Euclidean primal space of one layer. *CellSpace-Boundary* semantically describes the boundaries of each space object. Space features are further classified into two groups: *NavigableSpace* and *NavigableSpaceBoundary*. *NavigableSpace* represents all indoor spaces (e.g., rooms, corridors, doors, stairs) that can be used by a navigation application. *NavigableBoundary* represents the boundary of the spaces and is used to connect the navigation spaces (e.g., door).

The U3DSNM also can be mapped to another existing data model, CityGML. Figure 4.3 illustrates the UML diagram that maps the classes of the U3DSNM to classes of on-going version of CityGML 3.0. CityGML is an OGC standard, which is an open data model and XML-based format for the storage and exchange of virtual 3D city models. The aim of the development of CityGML is to reach a common definition of the basic entities, attributes, and relations of a 3D city model, in which spaces (class *AbstractSpace*) and space boundary (class *AbstractSpaceBoundary*) are introduced [172]. The *AbstractUnoccupiedSpace* means the unoccupied spaces that represent physical volumetric entities that do not occupy space in the urban environment, i.e., no space is blocked by these volumetric objects. Examples for unoccupied spaces are building rooms and traffic spaces. This definition aligns to the *Space* in U3DSNM. The class *ClosureSurface* represents the surfaces for enclosing an open space to volume. Therefore, the classes *Space*, *SpaceBoundary* in U3DSNM are mapped to the *AbstractUnoccupiedSpace*, *ClosureSurface* in CityGML 3.0 respectively.

4.5 DISCUSSION

The presented model is not restricted to the type of spaces and therefore can uniformly manage and represent the spaces for seamless navigation. Nevertheless, three aspects are worth discussing:

In this model, the accessibility of spaces is not considered. All the spaces are assumed available for navigation. However, under certain conditions or for specific users, spaces might be not accessible. For instance, some spaces are forbidden to be visited, because they are private, occupied, not allowed, or closed. The accessibility may be a critical characteristics mixed context of navigation. The current version of the model has no attribute that deals with this information. Further investigations are needed to reflect the accessibility of the four types of spaces in future work.

The model has been presented from a 3D point of view, but it remains valid if only 2D geometry of spaces is available. The navigation network will be still derived under unified Duality rules and will serve seamless navigation, but the height limitations of a path cannot be considered.

The model is based on 3D spaces, but such spaces might not be readily available and might need to be created. Currently, no data of software exist that can provide all four types of 3D spaces. Map providers or agencies cannot provide information on sO-spaces and sI-space, because they have to be closed with virtual boundaries. Even if BIM models and 2D maps exist, some 3D modeling process should take place. The data used in this research (Chapter 7) have been available from different sources, such as Estate Management, mapping agencies, or OpenStreetMap. If data about the environment is missing or outdated, algorithms for processing point clouds should be developed [149]. Fortunately, with the progress of 3D space modeling, we can access more 3D models, such as CityGML and IFC models. The on-going version of CityGML 3.0 is planning to model the outdoor as 3D spaces [172], which includes driving lanes, sidewalks, even waterway [188].

4.6 SUMMARY

This chapter presented a model U3DSNM, which contain information for space-based navigation in all kinds of environments. The model maintains I-spaces, sI-spaces, sO-spaces, and O-spaces and offers unified attributes, relationships, representation, management. The Duality theory ensures the navigation network extraction is standardized no matter the type of space. Importantly, this model can easily be mapped to existing data models such as the ongoing version of IndoorGML 2.0 and CityGML 3.0.

The presented model allows us to consider a larger variety of options for navigation using the classification of spaces. By assigning certain weights on spaces according to the users' preferences and tasks, it can help to achieve user-dedicated navigation paths, e.g., semantics-based path [189, 190]. Furthermore, the spaces can be enriched with many attributes and linked to supplementary data to derive user dedicated information. Such extended information can be used appropriately in many applications for navigation: shopping, facility management, guidance at airports and hospitals, etc.

Similar to the IndoorGML concepts, this model can support layers of spaces (subdivisions and aggregations), which can be organized in such a way to reflect user groups or individual users. For example, one layer of spaces in a campus area allows deriving a network for students, another for staff, and a third for facility managers. These layers could be then applied for location-based services (LBS) applications.

The experiments in this book are completed with a technical model maintained in the memory mentioned. However, the conceptual model (UML diagram) can be mapped to SQL implementation (spatial schema) and maintained in a database management system (DBMS). DBMS is recommendable for large data sets and frequent retrieval of data. Moreover, such implementation will ensure that many operations are completed at a database level and only the needed path is be sent to the end-user application. Many of the main-stream DBMS support network management and provide efficient routing operations.

The U3DSNM is important for urban data management and urban analytic as well. This model allows to consider and maintain the entire urban space/environment (indoor, outdoor, and transitional spaces). Many environmental characteristics (temperatures, noise, rain, pollution) can be attached to spaces and studies in greater depth, or combined studies on mobility and human activities. For example, people can use sI-spaces in case of rain or high temperatures [191]. Such studies can motivate urban planners and designers to consider more transitional spaces.

5 Three New Path Options

Currently, navigation path options in indoor and outdoor are monotonous as existing navigation systems commonly offer single-source shortest-distance or fastest paths. Such path options might be not always applicable. For instance, on a rainy day, a path with as many places that are covered by roofs/shelters might be preferable to the shortest path. Another example is that pedestrians in a shopping mall may be interested in a path that navigates through multiple places starting from and ending at the same location, i.e., a path similar to the traveling salesman problem (TSP) problem. A third example is when people may be unable to provide a specific destination. For example, a pedestrian wants to know which places he/she can reach within two minutes of walking from current location.

This chapter concentrates on the computation of navigation paths and presents three new path options, which are the Most-Top-Covered path (MTC-path), Path to the Nearest Semi-indoor (NSI-path), and Indoor Traveling Salesman Problem Path Planning (ITSP-path). The three path options make use of the space-based model introduced in the previous chapter. They are also intended to trigger new directions for designing navigation paths to enhance the navigation experience for pedestrians in indoor/outdoor space.

5.1 CURRENT RESEARCH ON NAVIGATION PATH

Current navigation systems primarily take the travel distance (shortest) or time (fastest) as the main criteria for optimal path computation. Such path options could be a habitual choice for humans [192], but it does not mean this kind of path is always the best choice under given circumstances.

In the past decades, except for the general shortest-distance/fastest path options, much attention has been paid to other navigation path options (Table 5.1). Furthermore, although various algorithms have been developed for path planning, such as the Dijkstra algorithm [193], A* [194], ant colony [195], Rapidly-exploring random tree (RRT) [196] and its variants [197]. To the best knowledge, the algorithms or approaches in the literature can be employed as basic algorithms but could not be directly utilized to handle indoor planning of MTC-path, NSI-path, and ITSP-path. For example, Dijkstra and A* are two commonly used algorithms, but they are only good at single-source shortest path planning, and need explicit departure and destination locations. In short, existing tools and algorithms become invalid for indoor new path options.

DOI: 10.1201/9781003281146-5

Table 5.1
Examples of indoor navigation path options.

Path Options	References
General safe path	[40, 41, 43]
Least or most-space-visited, least-obstruction	[87]
Simplest or minimum turns path	[198, 199]
Safe path avoiding dynamic obstacles	[200]
Health-optimal routing (e.g., a specific level of calories burn)	[201]
Minimum traffic-related air pollution exposure	[202]

5.2 TWO NEW SI-SPACE RELATED NAVIGATION PATH

5.2.1 PARAMETERS

In order to quantitatively reflect the navigation networks and navigation paths, this section introduces several parameters.

For Navigation Model The parameters for the navigation networks are defined based on two connected spaces (Figure 5.1). Only the two spaces (s_1 and s_2) in Figure 5.1(a) are regarded as connected spaces while the spaces (s_3 and s_4) in Figure 5.1(b) and (s_5 and s_6) in Figure 5.1(c) are unconnected. The Figure 5.1(d) shows the navigation network between s_1 and s_2, in which the blue dot is the additional vertex on the face two spaces touch to indicate how to traverse the spaces.

Five parameters are defined: the distance between the two connected spaces, original weights, covered & uncovered distance, uncovered ratio, and modified weights. The definitions and notations of the parameters are the following:

+ The distance between two connected spaces ($d_{s_i s_j}$)

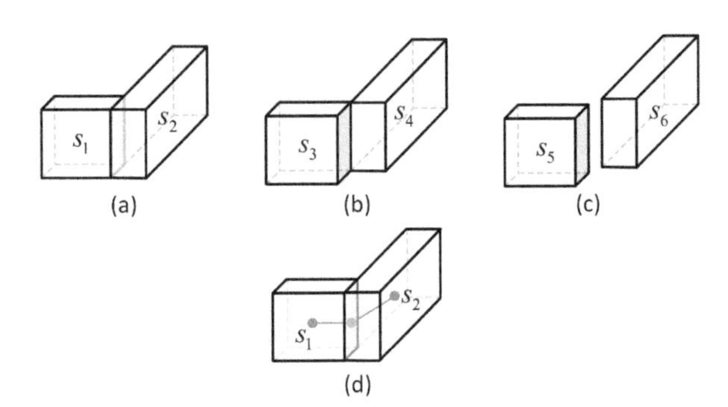

Figure 5.1 Illustration of connected spaces and the navigation network between the two spaces.

On the basis of Poincaré duality, for any two connected spaces (s_i and s_j), edge(s) are added to indicate connection(s), where the costs of edges are distances. Then, the distance between the two spaces ($d_{s_is_j}$) is the sum of all the costs between them. For instance, the distance between S_1 and S_2 is the sum of the lengths of the two line segments (Figure 5.1(d)).

+ Original weights ($W'_{s_is_j}$)
All the distances are taken into account to compute the weights. Then, the original weights are the standardized distances (Equation 5.1).

$$W'_{s_is_j} = \frac{d_{s_is_j} - d_{s_is_j}(min)}{d_{s_is_j}(max) - d_{s_is_j}(min)} \tag{5.1}$$

where $W'_{s_is_j}$ is the values that standardized from distances $d_{s_is_j}$, $d_{s_is_j}(max)$ and $d_{s_is_j}(min)$ are the maximum and minimum in the distance collection respectively.

+ Covered ($d_{c_{s_is_j}}$), & uncovered ($d_{uc_{s_is_j}}$) distance
The covered distance ($d_{c_{s_is_j}}$) means the length of the part physically bounded by tops between two connected spaces (s_i and s_j), and the uncovered distance ($d_{uc_{s_is_j}}$) means the length of the uncovered parts. In this research, the covered parts come from I-spaces or sI-spaces while the uncovered from sO-spaces or O-spaces.

+ Uncovered ratio ($\lambda_{s_is_j}$)
The uncovered ratio ($\lambda_{s_is_j}$) is a variable to express uncovered rate between two spaces. It is the ratio between the uncovered distance ($d_{uc_{s_is_j}}$) and the distance ($d_{s_is_j}$) of two spaces (s_i and s_j). Thus, it can be computed by Equation 5.2:

$$\lambda_{s_is_j} = d_{uc_{s_is_j}} / d_{s_is_j}. \tag{5.2}$$

where $d_{uc_{s_is_j}}$ is uncovered distance and $d_{s_is_j}$ is the distance between the two connected spaces.

+ Modified weights ($W''_{s_is_j}$)
The modified weights ($W''_{s_is_j}$) are computed based on original weights ($W'_{s_is_j}$) and the uncovered ratio ($\lambda_{s_is_j}$), in which a coefficient ξ is introduced to indicate the importance of the original weights and uncovered ratio (Equation 5.3).

$$W''_{s_is_j} = \xi W'_{s_is_j} + (1-\xi)\lambda_{s_is_j} \tag{5.3}$$

where ξ is the coefficient that quantifies the importance of original weights and uncovered ratio, where $\xi \in [0,1]$, $W'_{s_is_j}$ is the original weight, and $\lambda_{s_is_j}$ is the uncovered ratio.

For Navigation Path The parameters for the navigation paths are defined based on the planned paths, which are used as quantitative indicators for path comparisons in the later section. The definitions and notations of the parameters are as follows:

+ Path length (P_l)
  The P_l is the distance from departure to destination following the planned path.

$$P_l = \sum d_{s_i s_j} \tag{5.4}$$

+ Covered/Uncovered length of a path ($P_{l_c}/P_{l_{uc}}$)
  Covered length means the total distance of the path segments formed by I-spaces and sI-spaces in the planned path, while the uncovered distance means the total distance of segments formed by sO-space and/or O-space.

$$P_{l_c} = \sum d_{c_{s_i s_j}} \tag{5.5}$$

$$P_{l_{uc}} = \sum d_{uc_{s_i s_j}} \tag{5.6}$$

+ Top-coverage-ratio of a path (P_{c_r})
  The top-coverage-ratio of a path is an indicator that shows how much a path is physically bounded by tops. It is the ratio between the covered length and path length (Equation 5.7).

$$P_{c_r} = P_{l_c}/P_l \tag{5.7}$$

+ Total weight of a path (W_p)
  The total weight of a path is the summation of the weight of a planned path based on (original/modified) weights. In particular, if a planned path consists of several connected spaces, the W_p is the sum of weights corresponding to all the spaces (Equation 5.8).

$$W_p = \sum W_{s_i s_j} \tag{5.8}$$

in which the $W_{s_i s_j}$ is $W'_{s_i s_j}$ or $W''_{s_i s_j}$.

5.2.2 MTC-PATH

The MTC-path is a parameters-based path option and it takes both the travel distance and the top-coverage-ratio of a path (P_{c_r}) as the criteria. In short, it aims to determine the shortest path within a top-coverage-ratio constraint in the navigation. The path planning of MTC-path consists of three steps:

- Step 1: Select sI-spaces. In this step, sI-spaces will be selected based on a threshold of C^T. This threshold of C^T is different from that in the generic space definition framework. Here, it is set on basis of navigation purposes. For instance, if a pedestrian plans to use the top for escaping from strong sun, the threshold of C^T can be set as 0.8 (this value here is only an example). Then, only the sI-spaces with a $C^T \geq 0.8$ are selected to participate in the navigation network derivation and navigation path planning. In other words, the results of this step are the sI-spaces that have qualified tops.

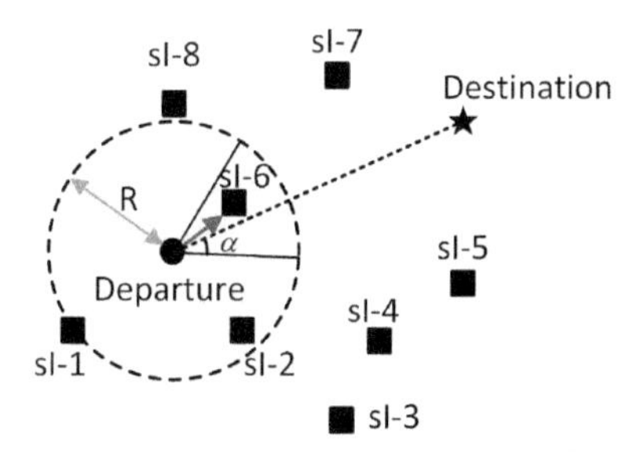

Figure 5.2 Example of NSI-path planning from departure to destination.

- Step 2: Compute the original and modified weights. Taking the Poincaré duality as the theoretical background, the navigation network is derived based on the selected sI-spaces, sO-spaces, and O-spaces. Then, the original and modified weights are computed based on the Equation 5.1 and Equation 5.3, respectively.
- Step 3: Plan the MTC-path. In this last step, the departure and destination are two nodes in the navigation network corresponding to two spaces and they can be located in semi-indoor, semi-outdoor, or outdoor. After assigning the departure and destination, the modified weights ($W''_{s_i s_j}$) are utilized to compute the MTC-path as well as P_l, P_{l_c}, $P_{l_{uc}}$.

5.2.3 NSI-PATH

The NSI-path is designed to help pedestrians to find a closest roofed/sheltered place from their departures, which is a compromise option when neither the shortest path nor MTC-path is recommended (see the path selection strategy in Section 5.2.4). The process of planning the NSI-path also starts from selecting sI-spaces based on a threshold of C^T (the same as that in MTC-path planning). The process of NSI-path planning consists of four steps, which are shown by an illustration (Figure 5.2). In the example, there are eight selected sI-spaces (sI-1 to sI-8) that are marked by black solid square. The details of the process are the following:

- Step 1: Create a straight line by linking the departure and destination. This step is to create a line segment by using the location of departure as the start and destination as the end. This line is used as a reference for the searching area determination.
- Step 2: Set time (t) and searching angle (θ). In this step, two parameters are introduced, time (t) and searching angle (θ), in which the former indicates the acceptable time for a pedestrian to move to the nearest sI-space, while the latter is an optimization parameter for determining the preferred search range. The θ can vary from 0 to 360°.
- Step 3: Find potential nearest sI-spaces. With the t and the speed of a pedestrian (v), a search radius (s_r) centered on the departure point can be

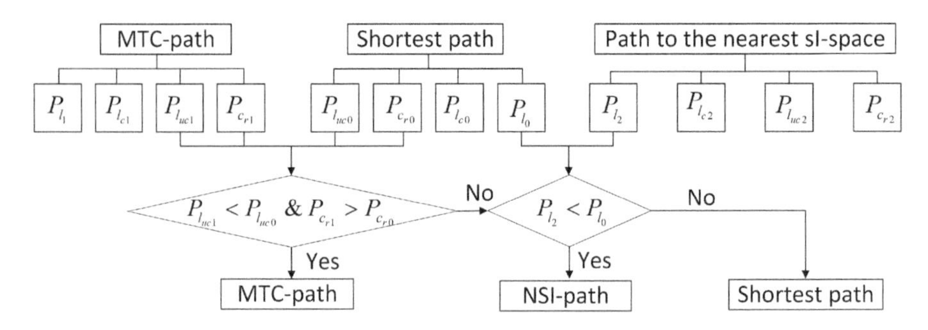

Figure 5.3 The path selection strategy. The shortest path is computed based on the original weights ($W'_{s_i s_j}$).

determined, i.e., $s_r = vt$. Then, the searching area becomes a sector by setting a searching angle ($\theta \in (0, 360^o]$). The searching process is having all the selected sI-spaces to do intersection operations with this sector. If the intersection is not null, the corresponding sI-space will be kept as a candidate for the nearest sI-spaces. With a given θ, if there is no sI-space within the defined sector, the θ will be increased.

- Step 4: Plan the nearest sI-space and NSI-path. The final step is computing the shortest paths from the departure to each candidate of sI-space based on the Dijkstra algorithm. Then the sI-space corresponding to the path with minimum distance is the nearest sI-space, and this shortest path is the NSI-path. In the demonstration (Figure 5.2), *sI-6* is determined as the nearest sI-space to the departure.

5.2.4 A PATH SELECTION STRATEGY

There are three options for a pedestrian, MTC-path, NSI-path, and the traditional shortest path. More than one paths options can be available and this may create difficulties in selecting a path. It is necessary to have a path selection strategy to help pedestrians to make decisions within the three path options.

Ideally, pedestrians will select the path options with the following order: MTC-path, NSI-path, and the traditional shortest path. However, MTC-path may not always be the best choice. For example, the length of MTC-path may be longer than the shortest path. Thence, this chapter introduces a path selection strategy (Figure 5.3) to help pedestrians to find the balance between distance and top-coverage-ratio and set up rules to estimate in which condition which path is the best option.

As seen in the path selection strategy, utilizing the traditional shortest path as the reference, two progressive rules are presented: (i) for the MTC-path, if its uncovered distance is shorter than that of shortest path, and at the same time the top-coverage-ratio is larger than that of shortest path (i.e., $P_{l_{uc1}} < P_{l_{uc0}}$ & $P_{c_{r1}} > P_{c_{r0}}$), it will be recommended to agents preferentially. Otherwise, (ii) the NSI-path will be computed and compared with the traditional shortest path. If the NSI-path is shorter than the

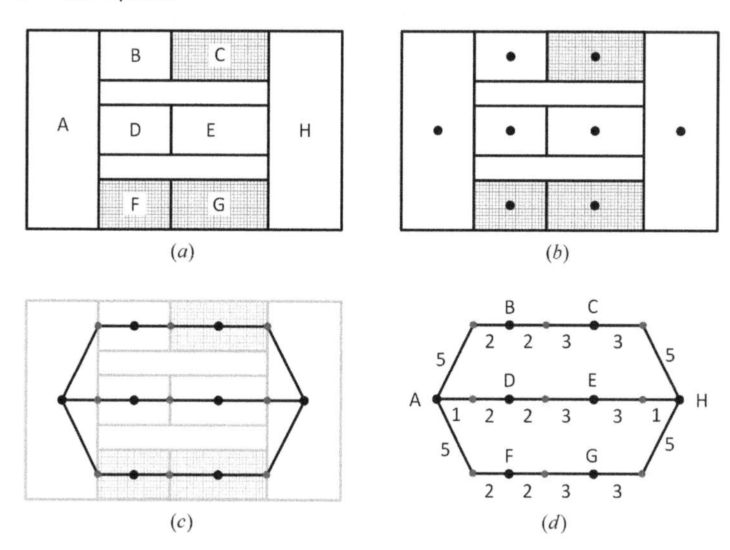

Figure 5.4 A navigation example, in which s_C, s_F, and s_G are three sI-spaces. The navigation graph in (c) and (d) are undirected graph. (a) All spaces. (b) Nodes extracted from spaces; (c) Navigation graph derived from spaces based on duality theory, in which the red dots are the extra vertices; (d) Navigation graph with distance.

shortest path (i.e., $P_{l_2} < P_{l_0}$), the NSI-path will be suggested for pedestrians. Otherwise, the traditional shortest path is the recommendation.

5.2.5 ILLUSTRATION OF THE TWO PATH OPTIONS

A navigation example illustrates the two navigation options and the path selection strategy (Figure 5.4). For simplicity, we use abstract cases containing only spaces which connect by sharing virtual boundaries. There are eight spaces (s_A to s_H), in which s_C, s_F, and s_G are three selected sI-spaces, while the rests are sO-spaces and O-spaces. Spaces are 3D volumes, but in this example, they are demonstrated in 2D polygons for visualization purposes. Figures 5.4(a) and (c) show the navigation network that derived based on Poincaré duality. The costs of edges are distances, which are identified by numbers without the unit (Figure 5.4(d)).

With the navigation graph (Figure 5.4(d)) and distances, original weights (Table 5.2) are computed based on Equations 5.2 and 5.1. For instance, the $W'_{s_A s_F} = (7-3)/(8-3) = 0.8$, in which $d_{s_A s_F} = 7$, $d_{s_i s_j}(min) = min\{7,3,7,5,8,5,4,5,8\} = 3$ while $d_{s_i s_j}(max) = min\{7,3,7,5,8,5,4,5,8\} = 8$. The uncovered ratio between s_A and s_F is $\lambda_{s_A s_F} = d_{uc_{s_i s_j}}/d_{s_i s_j} = 5/7 = 0.71$. Then, modified weights ($W''_{s_i s_j}$) are computed based on Equation 5.3, in which to show the changes in the modified weights, the coefficient ξ is set from 1 to 0 with intervals of 0.1 (Table 5.3). For example, when $\xi = 0.6$, the modified weight of the link between space s_A and s_F is $W''_{s_A s_F} = \xi W'_{s_A s_F} + (1-\xi)\lambda_{s_A s_F} = 0.6 * 0.8 + (1-0.6) * 0.71 = 0.77$.

Table 5.2
Original information of the navigation graph.

s_i	s_j	$d_{s_is_j}$	$W'_{s_is_j}$	$d_{c_{s_is_j}}$	$d_{uc_{s_is_j}}$	$\lambda_{s_is_j}$
s_A	s_B	7	0.8	0	7	1
s_A	s_D	3	0	0	3	1
s_A	s_F	7	0.8	2	5	0.71
s_B	s_C	5	0.4	3	2	0.4
s_C	s_H	8	1	3	5	0.625
s_D	s_E	5	0.4	0	5	1
s_E	s_H	4	0.2	0	4	1
s_F	s_G	5	0.4	5	0	0
s_G	s_H	8	1	3	5	0.625

Table 5.3
Modified weights ($W''_{s_is_j}$) based on Equation 5.3.

s_i	s_j	$W''_{s_is_j}$										
		$\xi = 1$	0.9	0.8	0.7	0.6	0.5	0.4	0.3	0.2	0.1	0
s_A	s_B	0.8	0.82	0.84	0.86	0.88	0.9	0.92	0.94	0.96	0.98	1
s_A	s_D	0	0.1	0.2	0.3	0.4	0.5	0.6	0.7	0.8	0.9	1
s_A	s_F	0.8	0.79	0.78	0.77	0.77	0.76	0.75	0.74	0.73	0.72	0.71
s_B	s_C	0.4	0.4	0.4	0.4	0.4	0.4	0.4	0.4	0.4	0.4	0.4
s_C	s_H	1	0.96	0.93	0.89	0.85	0.81	0.78	0.74	0.7	0.66	0.63
s_D	s_E	0.4	0.46	0.52	0.58	0.64	0.7	0.76	0.82	0.88	0.94	1
s_E	s_H	0.2	0.28	0.36	0.44	0.52	0.6	0.68	0.76	0.84	0.92	1
s_F	s_G	0.4	0.36	0.32	0.28	0.24	0.2	0.16	0.12	0.08	0.04	0
s_G	s_H	1	0.96	0.93	0.89	0.85	0.81	0.78	0.74	0.70	0.66	0.63

With the navigation network and weights, three navigation paths from s_A (departure) to s_H (destination) are planned (Figure 5.5), in which path 1 ($s_A \rightarrow s_D \rightarrow s_E \rightarrow s_H$) is the traditional shortest path, path 2 ($s_A \rightarrow s_F \rightarrow s_G \rightarrow s_H$) is the MTC-path. The path 3 ($s_A \rightarrow s_F$) is the NSI-path. It should be clarified that in this case, the path 3 is a part of the path 2, because it happens that s_F is the sI-space closest to the departure s_A. Nevertheless, if the closest sI-space to s_A is another space rather than the s_F, this coincidence will disappear.

The information of the three path options are listed in Table 5.4. To show how the information are computed, we select $1-\xi = 0.3$ (i.e., $\xi = 0.7$) as an example. For the path 1, $W_p = W''_{s_As_D} + W''_{s_Ds_E} + W''_{s_Es_H} = 0.3 + 0.58 + 0.44 = 1.32$, $P_l = d_{s_As_D} + d_{s_Ds_E} +$

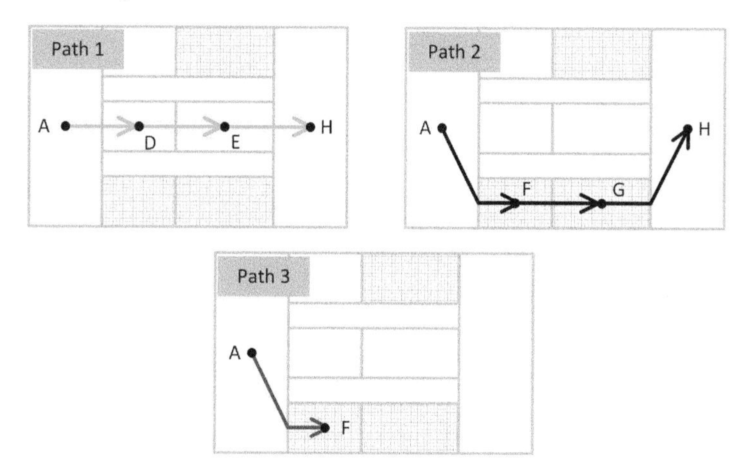

Figure 5.5 The three navigation paths from s_A (departure) to s_H (destination). $s_A \to s_D \to s_E \to s_H$ is path 1 (green), $s_A \to s_F \to s_G \to s_H$ is path 2 (black), and $s_A \to s_F$ is path 3 (blue).

Table 5.4
Three different navigation paths.

$1-\xi$	Path 1				Path 2				Path 3			
	W_p	P_l	$P_{l_{uc}}$	P_{c_r}	W_p	P_l	$P_{l_{uc}}$	P_{c_r}	W_p	P_l	$P_{l_{uc}}$	P_{c_r}
0	0.6	12	12	0	2.2	20	10	0.5	0.8	7	5	0.29
0.1	0.84	12	12	0	2.11	20	10	0.5	0.79	7	5	0.29
0.2	1.08	12	12	0	2.03	20	10	0.5	0.78	7	5	0.29
0.3	1.32	12	12	0	1.94	20	10	0.5	0.77	7	5	0.29
0.4	1.56	12	12	0	1.86	20	10	0.5	0.77	7	5	0.29
0.5	1.8	12	12	0	1.77	20	10	0.5	0.76	7	5	0.29
0.6	2.04	12	12	0	1.68	20	10	0.5	0.75	7	5	0.29
0.7	2.28	12	12	0	1.6	20	10	0.5	0.74	7	5	0.29
0.8	2.52	12	12	0	1.51	20	10	0.5	0.73	7	5	0.29
0.9	2.76	12	12	0	1.43	20	10	0.5	0.72	7	5	0.29
1	3	12	12	0	1.34	20	10	0.5	0.71	7	5	0.29

$d_{s_E s_H} = 3+5+4 = 12$, $P_{l_{uc}} = d_{uc_{s_A s_D}} + d_{uc_{s_D s_E}} + d_{uc_{s_E s_H}} = 3+5+4 = 12$, and $P_{c_r} = (P_l - P_{l_{uc}})/P_l = (12-12)/12 = 0$. Similarly, for the path 2, $W_p = W''_{s_A s_F} + W''_{s_F s_G} + W''_{s_G s_H} = 0.77+0.28+0.89 = 1.94$, $P_l = d_{s_A s_F} + d_{s_F s_G} + d_{s_G s_H} = 7+5+8 = 20$, $P_{l_{uc}} = d_{uc_{s_A s_F}} + d_{uc_{s_F s_G}} + d_{uc_{s_G s_H}} = 5+0+5 = 10$, and $P_{c_r} = (P_l - P_{l_{uc}})/P_l = (20-10)/20 = 0.5$. As for the path 3, $W_p = w''_{s_A s_F} = 0.77$, $P_l = d_{s_A s_F} = 7$, $P_{l_{uc}} = d_{uc_{s_A s_F}} = 5$, and $P_{c_r} = (P_l - P_{l_{uc}})/P_l = (7-5)/7 = 0.29$.

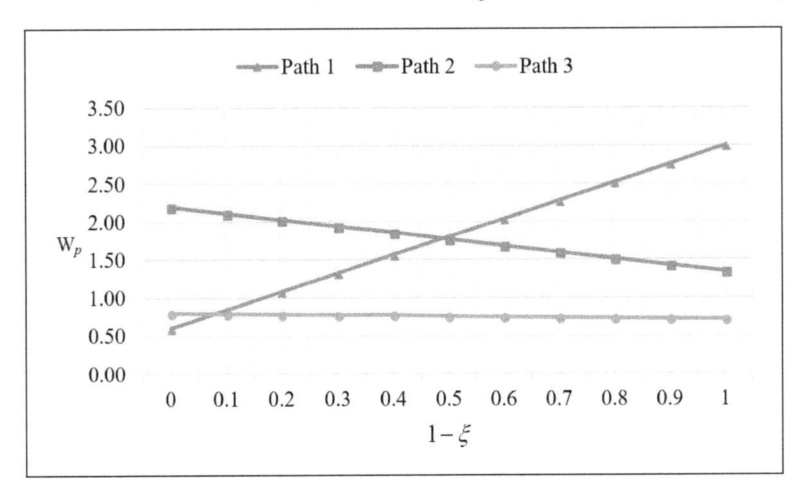

Figure 5.6 The changes of W_p with the changing of the ξ.

It shows that with the changing of the ξ, the W_p of the three navigation paths change (Figure 5.6). Overall, with the decreasing of ξ (i.e., the increasing of $1-\xi$), the W_p of path 1 is rising, while that of path 2 and 3 are falling, which reveals that with paying more attention to the top-coverage-ratio of the path, the traditional shortest path becomes less attractive. That is, if ξ is less than 0.5, the path 2 is recommended, otherwise, the recommendation becomes the path 3.

Comparing paths 1 and 2, before ξ reaching to 0.4, the path 2 is recommended for agents. For instance, if $\xi = 0.4$, The uncovered distance of path 2 ($P_{l_{uc2}} = 12$) is shorter than that of path 1 ($P_{l_{uc1}} = 14$), although W_p of path 2 is smaller than that of path 1, and the top-coverage-ratio ($P_{c_{r2}} = 0$) is smaller than that of path 1 ($P_{c_{r1}} = 0.3$). For this case, NSI-path is computed. But, path 2 is still be recommended, because the length of NSI-path is longer than that of path 2.

In contrast, comparing the path 2 and path 3, we can find that if ξ less than to 0.5, the path 2 is recommended, otherwise, the recommendation is the path 3. Because when the $\xi > 0.5$, the W_p of path 3 smaller than that of path 2, the uncovered distance of path 3 is shorter that than of path 2, and the top-coverage-ratio changes larger than that of path 2.

Comparing paths 1 and 3 only, at the beginning (i.e., $\xi = 1$), the two paths are the same from the W_p aspect. But, with the decreasing of (ξ), path 3 becomes the recommended path considering its W_p becomes always less than that of path 1. Furthermore, its covered distance is longer than that of path 1 ($P_{l_c3} > P_{l_c1}$) and top-coverage-ratio is bigger than that of path 1 ($P_{c_{r3}} > P_{c_{r1}}$).

5.3 ITSP-PATH

As mentioned above, some pedestrians may be interested in the shortest path that can navigate them from the departure and come back to the original point after

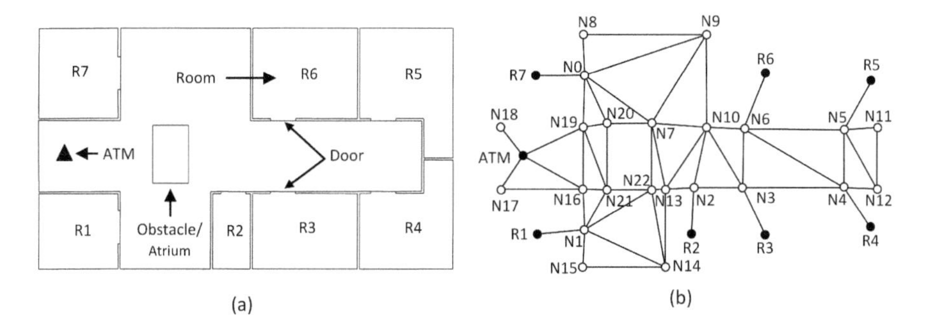

Figure 5.7 An indoor map and its navigation network.

visiting specific places. This is a very typical case for pedestrians visiting shopping malls of shopping streets. In this book, such kind of shortest path is named as indoor travelling salesman problem (ITSP) path to acknowledge that it is a variation of the conventional Travelling Salesman Problem (TSP). The conventional TSP is a well-known algorithmic problem in the fields of computer science and operations research, which was defined as the challenge of finding the shortest, yet most efficient route for a person (or vehicle) to follow, given a list of specific destinations [203]. In short, TSP is a way of a movement when the traveller must visit several destinations only once and return to the origin.

The ITSP may the same as the conventional TSP, however, it is not identical. In the conventional TSP, all the nodes are destinations and all of them will be included in the navigation path, while the navigation path of ITSP may include only a part of all possible nodes. In other words, if there are n nodes in the conventional TSP graph, the navigation path for this case will includes all the nodes. Yet, if there are n nodes in navigation network for ITSP, the navigation path will include m ($m \leq n$) nodes. This is to say, the conventional TSP should be applied at a different set of nodes each time a navigation is requested.

We illustrate the difference between conventional TSP and ITSP with an indoor example. Figure 5.7 shows an indoor map and corresponding indoor navigation network. There are seven rooms (R1 - R7), one Point of Interest (POI) - Automated Teller Machine (ATM) (the black dots in the navigation network), and one obstacle/atrium. The rooms and POI could be desired destinations. The other twenty three nodes (N0 - N22) marked by hollow circles are other indoor locations, here, we name them as dummy nodes. If a person desires to start from R7 and wants to come back after visiting ATM, R1, R2, and R5, the conventional TSP will fail to compute the shortest path, because it generally would include all thirty nodes. Such path is not reasonable for this indoor application, since one shortest path that contains the above five desired nodes (R7, ATM, R1, R2, and R5) at once and forms a ring-like path is what the pedestrian needs. Therefore, we argue that it is necessary to adapt the conventional TSP for indoor applications. This book proposes a solution to ITSP path planning by combining Dijkstra algorithm [193] and branch and bound (B&B) algorithm [204, 205].

5.3.1 CONCEPTS AND MODELING

The ITSP can be best described as the following scenario: a pedestrian plans to visit an indoor scene with several ($n \in Z^+$) indoor nodes (they could be rooms, indoor POIs, specific functional areas, etc.), in which the distance between every two nodes ($d(s_i, s_j)$) may be the same or different. The pedestrian starts from a space and return to the same space after visiting all the desired indoor nodes. The solution to this problem is to seek the optimal order of visiting all desired indoor nodes that makes the path to be the shortest.

The mathematical model of ITSP is:

All indoor spaces are represented as nodes:

$$S = \{s_1, s_2, s_3, ...s_n\} \tag{5.9}$$

The ITSP is symmetric [206], the distances between any two nodes are:

$$d(s_i, s_j) = d(s_j, s_i) \in R^+, 1 \leq i < j \leq n \tag{5.10}$$

where i and j mean the ith and jth of the indoor nodes.

The goal is to find the optimal order of visiting all desired indoor nodes to minimize the cost:

$$min\{\sum_{i=1}^{n-1} d(s_{k_i}, s_{k_{i+1}}) + d(s_{k_n}, s_{k_1})\} \tag{5.11}$$

where s_{k_i} denotes the k_ith indoor node.

5.3.2 PROCEDURES OF ITSP-PATH PLANNING

The path planning of ITSP is based on an indoor navigation network, which is modeled as a graph ($G_{original}(V,E)$). In the graph, vertices (V) are abstracted from indoor spaces s_i and edges (E) from the relationships between spaces (the theoretical basis is Poincaré duality [150]). After choosing one node as the departure location ($s_{departure}$) and several other nodes (all the other nodes can be included in a set $s_{intermediate}$) as specified intermediate destinations, a shortest path that includes all intermediate destinations once and go back to the destination ($s_{departure}$) is needed to be computed. The procedures of ITSP path planning include the following five steps:

- Step 1: Select intermediate destinations (nodes);
  Select departure location ($s_{departure}$) and specified intermediate places ($s_{intermediate}$). For instance, the $s_{departure} = s_1$ and $s_{intermediate} = \{s_4, s_7, s_9\}$.
- Step 2: Compute navigation paths between every two selected nodes;
  This step takes departure location and specified intermediate places as nodes to compute navigation paths between any two nodes based on Dijkstra. The lengths of the navigation paths are used as the distances between two nodes. In this case, we set the $d(s_i, s_j) = d(s_j, s_i)$, i.e., the paths in this example are symmetric and we simply assume the route between

Table 5.5
Example of the undirected graph.

	s_1	s_4	s_7	s_9
s_1	0	$d(s_1, s_4)$	$d(s_1, s_7)$	$d(s_1, s_9)$
s_4	$d(s_4, s_1)$	0	$d(s_4, s_7)$	$d(s_4, s_9)$
s_7	$d(s_7, s_1)$	$d(s_7, s_4)$	0	$d(s_7, s_9)$
s_9	$d(s_9, s_1)$	$d(s_9, s_4)$	$d(s_9, s_7)$	0

the two intermediate destinations is bi-directional. In reality, $d(s_i, s_j)$ may be not equal to $d(s_j, s_i)$, e.g., one-way corridors such as escalators. In all cases, $d(s_i, s_i) = d(s_j, s_j) = 0$. Continue the example in Step 1, this step can get six paths: $s_1 \rightsquigarrow s_4$, $s_1 \rightsquigarrow s_7$, $s_1 \rightsquigarrow s_9$, $s_4 \rightsquigarrow s_7$, $s_4 \rightsquigarrow s_9$, $s_7 \rightsquigarrow s_9$, and corresponding reverse paths. The symbol ($\rightsquigarrow$) means there are zero to several navigation nodes between two nodes. The travel distances are $d(s_1, s_4) = d(s_4, s_1)$, $d(s_1, s_7) = d(s_7, s_1)$, $d(s_1, s_9) = d(s_9, s_1)$, $d(s_4, s_7) = d(s_7, s_4)$, $d(s_4, s_9) = d(s_9, s_4)$, $d(s_7, s_9) = d(s_9, s_7)$.

- Step 3: Set up graph of desired intermediate destinations;
  Setting up graph of all desired intermediate destinations is to take the travel distances of every two specified places as weights and all these places as nodes to make an undirected graph. For the example in Step 1, the undirected graph can be organized as a table (Table 5.5).
- Step 4: Select and sort the navigation paths based on B&B algorithm;
  Taking the undirected graph as the input, the orders of departure and desired intermediate destinations can be computed based on the B&B algorithm. Then, the orders are further used to select and sort navigation paths. For instance, the order could be $< s_1, s_7, s_9, s_4, s_1 >$, which means the following navigation paths will selected and sorted as: $s_1 \rightsquigarrow s_7$, $s_7 \rightsquigarrow s_9$, $s_9 \rightsquigarrow s_4$, and $s_4 \rightsquigarrow s_1$.
- Step 5: Combine the navigation results of Dijkstra as the ITSP-path.
  The last step is to combine the navigation results as the ITSP-path. For instance, the ITSP-path of the example is: $s_1 \rightsquigarrow s_7 \rightsquigarrow s_9 \rightsquigarrow s_4 \rightsquigarrow s_1$. Then, the final navigation path becomes: $s_1 \rightarrow ... \rightarrow s_7 \rightarrow ... \rightarrow s_9 \rightarrow ... \rightarrow s_4 \rightarrow ... \rightarrow s_1$. The symbol ($\rightarrow$) means there is no other navigation node between the two nodes.

5.3.3 ILLUSTRATION

An indoor example is employed to demonstrates the procedures of ITSP-path planning. The navigation network of the indoor scenario is shown in Figure 5.8(a). In the first step, we select the room (“R7”) as departure location and specified intermediate

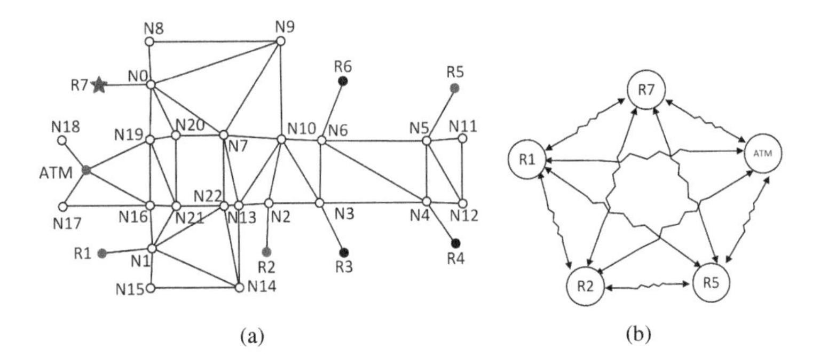

Figure 5.8 (a) The navigation network of the indoor scene; (b) Illustration of the undirected graph of all selected desired destinations.

Table 5.6
Navigation path between any two selected places.

Start	End	Path	Distance
R7	R1	R7 → N0 → N19 → N16 → N1 → R1	12.19
R7	R2	R7 → N0 → N7 → N13 → N2 → R2	13.24
R7	R5	R7 → N0 → N7 → N10 → N6 → N5 → R5	18.43
R7	ATM	R7 → N0 → N19 → ATM	8.19
R1	R2	R1 → N1 → N22 → N13 → N2 → R2	10.45
R1	R5	R1 → N1 → N22 → N13 → N10 → N6 → N5 → R5	19.81
R1	ATM	R1 → N1 → N16 → ATM	7.70
R2	R5	R2 → N2 → N10 → N6 → N5 → R5	14.50
R2	ATM	R2 → N2 → N13 → N22 → N21 → N16 → ATM	11.04
R5	ATM	R5 → N5 → N6 → N10 → N7 → N20 → N19 → ATM	18.73

places are “R1”, “R2”, “R5”, “ATM”, i.e., $s_{departure}$ =“R7” (marked by red star) and $s_{intermediate}$ = {“$R1$”,“$R2$”,“$R5$”,“ATM”} (marked by blue dots).

By following the second step in the procedures, navigation paths of every two desired destinations can be computed based on Dijkstra. Utilizing each nodes as the temporary departure and destination, we can get all the navigation paths and travel distances (Table 5.6). Then, the distances are utilized to make up an undirected graph (Figure 5.8(b) and Table 5.7).

With the undirected graph, the optimal orders of the nodes can be computed by using B&B algorithm: <“R7”, “R5”, “R2”, “R1”, “ATM”, “R7”>. It means that, having “R7” as the departure and “R1”, “R2”, “R5”, “ATM’ as the specified intermediate places during the traveling, the sorted navigation path is: R7 ⇝ R5 ⇝ R2 ⇝ R1 ⇝ ATM ⇝ R7 (Figure 5.9(a)).

Table 5.7
Undirected graph of all selected places.

	R7	R1	R2	R5	ATM
R7	0	12.19	13.24	18.43	8.19
R1	12.19	0	10.45	19.81	7.70
R2	13.24	10.45	0	14.50	11.04
R5	18.43	19.81	14.50	0	18.73
ATM	8.19	7.70	11.04	18.73	0

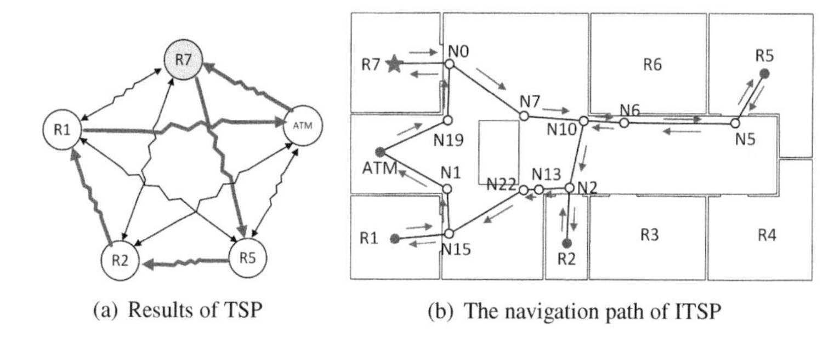

(a) Results of TSP (b) The navigation path of ITSP

Figure 5.9 The result of TSP and the optimal navigation path of ITSP.

The last step is to combine the results of Dijkstra to the sorted path as the ITSP-path. For the case, the optimal path is: R7 → N0 → N7 → N10 → N6 → N5 → R5 → N5 → N6 → N10 → N2 → R2 → N2 → N13 → N22 → N1 → R1 → N1 → N16 → ATM → N19 → N0 → R7 (Figure 5.9(b)). The cost of this path is 59.27.

5.4 SUMMARY

Three path options based on the different spaces are illustrated in this chapter. The first two utilize sI-spaces, including: MTC-path and NSI-path. Both paths are developed to meet the user's pursuit of protection from the top direction during their navigation.

The ITSP-pah can deal with the situation that pedestrians may be interested in visiting several locations and coming back to the starting point considering the shortest/fastest way. Although the path is demonstrated for indoor, it can be applied for many other indoor, outdoor or mixed scenarios by utilizing the U3DSNM. A museum is another indoor example, where the ITSP-path can assist in visiting only the top twenty exhibits or those organized in specific themes. Hospital visitors may need to visit multiple locations for examinations distributed in different buildings, which will require use of sI and sO spaces.

6 Reconstruction of 3D Navigation Spaces

On the basis of 3D spaces and Poincaré duality, navigation networks can be automatically derived. Nevertheless, 3D space models of the four types of spaces are not always available. No existing application can offer 3D space models of semi-indoor, semi-outdoor and outdoor, because the three types of spaces are partially or entirely open. BIM models can be data sources of indoor 3D space models, but the problem is BIM models are not always handy, such as some old built buildings have no such kind of data.

This chapter presents the approaches for reconstructing different types of spaces as *CellSpace*, in which the 3D geometry of a *CellSpace* is a volume that are combined of several *CellSpaceBoundary* (surfaces). That is, the 3D *CellSpace* reconstruction is a process of automatically reconstructing *CellSpaceBoundary* and joining them into 3D volumes. This introduced approaches can automatically reconstruct indoor, semi-indoor, semi-outdoor, and outdoor spaces as 3D spaces (volume). On the basis the approaches, all four types of spaces can mimic the indoor environments to derive a network based on the 3D connectivity of spaces. Four types of spaces are reconstructed for the purpose of testing the developed theories and models in this research. Considering all possible data sources, the reconstructing approach of the different spaces are the following:

(a) sI-spaces. The approach of sI-spaces reconstruction is creating 3D spaces base on projection operations between components that inspired from CityGML LoD 3 models, such as *RoofSurface*, *GroundSurface*, *OuterCeilingSurface*, and *OuterFloorSurface*.
(b) sO-spaces and O-spaces. The approach of reconstructing the sO-spaces and O-spaces is enclosing them into 3D enclosed volumes based on their footprints, a contact height, and DTM.
(c) Building shells. The approach of reconstructing the building shells is extruding their footprints based on building heights and DTM.

6.1 SEMI-INDOOR SPACE RECONSTRUCTION

The sI-space reconstruction is a process of enclosing sI-spaces as 3D enclosed volumes by creating missing sides. Taking advanced level of detail models (e.g., minimum CityGML LoD3 or BIM) as the data sources, the reconstruction includes the following steps:

- Step 1: Identification & ordering of proper building components;
- Step 2: Determination of top and bottom & space generation;
- Step 3: Space trimming.

DOI: 10.1201/9781003281146-6

6.1.1 IDENTIFICATION & ORDERING OF PROPER BUILDING COMPONENTS

Built structures, such as balconies, dormers or outer stairs could form sI-spaces. Therefore, the first step is to detect such building components in the input 3D models. The ground is also necessary, especially in cases that have no floors, e.g., a gas station. For this purpose, we strongly rely on the semantic information provided in the model and we assume advanced level of detail models (e.g., minimum CityGML LoD3 or BIM) are ready and the building components are identified based on their semantics.

Then these identified building components are sorted on the basis of their average value of the z coordinates. A virtual geometric *XY* plane is considered as the reference and its z coordinates are the minimum among all components. If more than one of the components have the same value of the z, they are ordered randomly.

The average value of the z coordinate of each component is computed using Equation 6.1:

$$z_A = \frac{1}{n}\sum_{i=1}^{n} z_i \tag{6.1}$$

where z_A = the average of the z coordinate of the component A
n = number of vertices of component A
z_i = the z coordinate of i_{th} vertex

6.1.2 DETERMINATION OF TOP AND BOTTOM & SPACE GENERATION

From higher to lower components (in the direction of the *Z-direction*), based on projection, overlaps that symbolize the desired space boundaries are detected. If the building components are volumes (e.g., in BIM models), a pre-processing step to this one includes a decomposition of the volumes into relevant upper and lower polygons. The process of space reconstruction between any two polygons includes four steps: (i) project the original two polygons onto the same plane along the *Z-direction*; (ii) compute the overlaps between their projections by intersection; (iii) find the tops and bottoms by projecting the overlaps back onto the two original polygons along the *Z-direction*; and (iv) obtain the missing lateral sides that allows forming closed volumes. In the second step, the projections on the same plane lead to one of two possible spatial relationships: either they overlap, or they do not. If their projections overlap, the region of intersection will be computed. Moreover, if the overlap is not a polygon (a line or point, for example), it is regarded as a case of no overlap.

Figure 6.1 illustrates the determination of top(s) and bottom(s). Then, the reconstruction of sI-spaces includes four different cases. The original polygons are A_i and B_i. Assuming their projections are A'_i and B'_i, respectively, the relationships of their projections are $A'_1 = B'_1$, $A'_2 \subset B'_2$, $B'_3 \subset A'_3$, and $A'_4 \cap B'_4 \neq \emptyset$ (sub-figures in the first column). Then, the overlaps of their projections (I_{AB}) are computed by intersection; i.e., $I_{A'_1B'_1} = A'_1(B'_1)$, $I_{A'_2B'_2} = A'_2$, $I_{A'_3B'_3} = B'_3$, and $I_{A'_4B'_4} = A'_4 \cap B'_4$ (sub-figures in the second column). The third step is to project the overlaps ($I_{A'_1B'_1}$, $I_{A'_2B'_2}$, $I_{A'_3B'_3}$, and

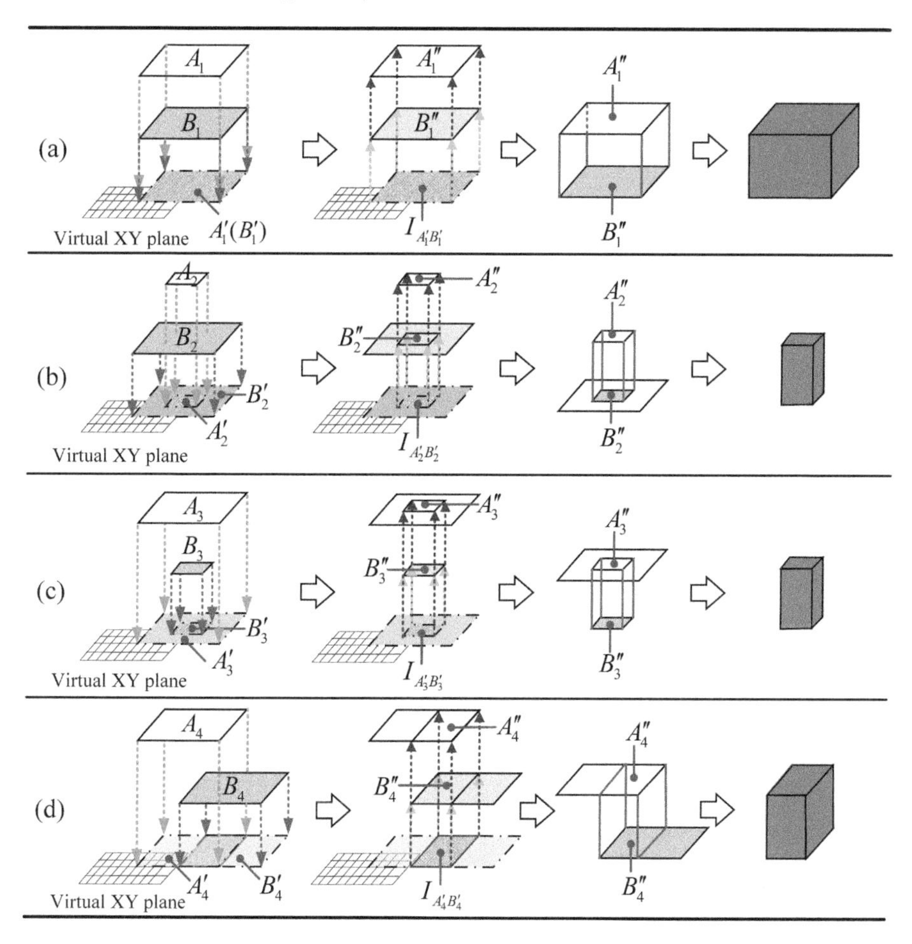

Figure 6.1 Four different cases of sI-space reconstruction based on projecting. The original polygons are A_i (in white) and B_i (in yellow), and their projections are A_i' (in gray) and B_i' (in light green) respectively. The created sI-spaces are colored in dark blue.

$I_{A_4'B_4'}$) back to the original polygons along the *Z-direction*; then, top(s) and bottom(s) are calculated: $\{A_1'', B_1''\}$, $\{A_2'', B_2''\}$, $\{A_3'', B_3''\}$, and $\{A_4'', B_4''\}$ (sub-figures in the third column). A_i'' and B_i'' are parts of polygons A_i and B_i, respectively, acting as top and bottom. Projections of A_i'' and B_i'' are equal to $A_i' \cap B_i'$. Because our approach relies entirely on the *Z-direction* as the projection direction, the last step is to match the top and bottom directly to obtain the missing lateral polygons (sides) that allow forming closed volumes (sub-figures in the last column). In addition, these operations are applicable to polygons that are not horizontal (inclined cases), but the vertical cases are not, as the projections of vertical components on the *Z-direction* are lines rather than polygons.

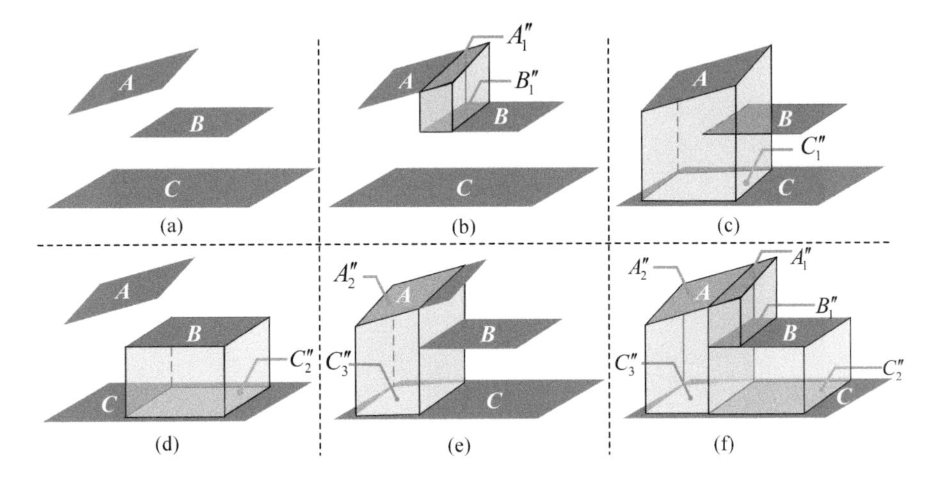

Figure 6.2 Example of trimming spaces. To distinguish sI-spaces that are created based on different tops and bottoms, they are colored differently. (a) shows the three building components; (b), (c), and (d) are three sI-spaces based on projections; (e) is the space trimmed by the other two spaces; (f) the final three reconstructed sI-spaces.

For the whole procedure of reconstructing all sI-spaces, any two polygons in the ordered component set are chosen to repeat the four steps. Then, all sI-spaces are reconstructed as 3D spaces.

6.1.3 SPACE TRIMMING

The final step is to trim the reconstructed 3D sI-spaces (volumes) based on the their positions, because these spaces may have overlapping parts.

Figure 6.2 illustrates the space reconstruction and trimming among three building components, in which the B and C are horizontal, while A is inclined. Suppose A is an inclined roof, B is a floor slab, and C is the ground. Repeating the four steps described in the last section three times, three sI-spaces are reconstructed, which tops and bottoms are $\{A''_1, B''_1\}$ (Figure 6.2(b)), $\{A, C''_1\}$ (Figure 6.2(c)), and $\{B, C''_2\}$ (Figure 6.2(d)). The space that top and bottom is $\{A, C'_1\}$ is overlapping with the other two. Thus, it is trimmed by the spaces shown in Figures 6.2(b) and 6.2(d). The trimmed result can be seen in Figure 6.2(e), in which A''_2, and C''_3 are the top and bottom, respectively. Thus, for this example, the final results of sI-spaces reconstruction are three spaces without overlaps, in which top(s) and side(s) are $\{A''_1, B''_1\}$, $\{A''_2, C''_3\}$, and $\{B, C''_2\}$ (Figure 6.2(f)).

6.1.4 ILLUSTRATION

The most of the cases where sI-spaces can be detected will likely appear in CityGML LoD3 (and above) models or BIM models. However, due to the lack of availability of LoD3 or LoD4 data, this research manually created a synthetic LoD3 model for

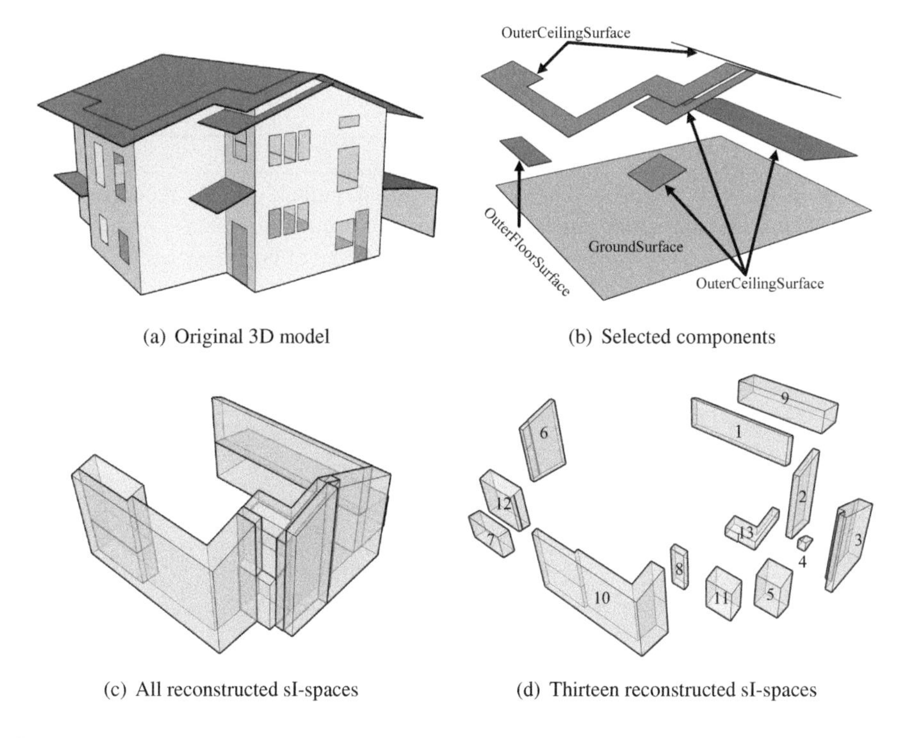

(a) Original 3D model
(b) Selected components
(c) All reconstructed sI-spaces
(d) Thirteen reconstructed sI-spaces

Figure 6.3 Illustration of sI-space reconstruction.

the sake of illustration of the approach. It is a two-story house with inclined roofs, a balcony, a shelter, and a garage (Figure 6.3(a)). First of all, we selected all the *OuterCellingSurface* and *OuterFloorSurface*. In the meantime, we used a default plane polygon extended from the floor polygon as the *GroundSurface* (Figure 6.3(b)). Secondly, all the sI-spaces are reconstructed (Figure 6.3(c)). After trimming spaces, we finally obtained thirteen sI-spaces (Figure 6.3(d)).

It is worth mentioning that although the illustration is a house, it is already sufficiently complex to be able to represent real-world cases. Thence, with this approach, the sI-spaces of the built structures (e.g., gas station, bus stop) can be reconstructed. Their roofs (covers/shelters) are used as the tops, and the *GroundSurface* as bottom. Another thing should be mentioned is that the CityGML LoD3 models or BIM models are the bottleneck of this approach.

6.1.5 ALGORITHMS

Based on the relationships of two surfaces, the upper and lower boundaries of all the spaces can be determined by Algorithm 1, in which, F'_i and F'_j are the areas in surface F_i and F_j, which projections are equal to $F_i \cap F_j$. The F'_i can be calculated by projecting the overlap of their projection onto F_i along the *Z-direction*, so does F'_j. The output of this algorithm is a list of boundary combinations B. Each element

$B[i]$ in B is a three-tuple, e.g., $B[i] = \{i, j, \{U, L\}\}$, in which the i is the Id of the upper boundary, and the j is the Id of the lower boundary. For the ground, this research uses the $'Ground'$ directly as a Id. $\{U, L\}$ is the geometry pair of the surfaces i and j.

Algorithm 1 Tops and bottoms determination.

Input: F, G ▷ F is a list of sorted surfaces based on their average Z coordinate, in which each component (surface) is F_i. G is the default ground.

Output: B ▷ A list of boundary combinations B

```
1:  procedure BOUNDARY DETERMINATION(F,G)
2:      U = F0                                              ▷ U is the top
3:      L = G                                               ▷ L is the bottom
4:      temp = 0
5:      for each Fi in S do
6:          temp = temp + 1
7:          projection1 = ProjectAlong(Fi)
8:          for each Fj in S (with j = i + 1) do
9:              if j == Size of S then
10:                 GProjection = ProjectAlong(Ground)
11:                 RInt = RegionIntersection(projection1, GProjection)
12:                 G' = ProjectOnSurface(RInt, Ground)
13:                 L = G'
14:                 Add the pair [i,'Ground',{U,L}] in B[i]
15:             else
16:                 projection2 = ProjectAlong(Fj)
17:                 RInt = RegionIntersection(projection1, projection2)
18:                 if RInt then                ▷ if the intersection is not null
19:                     F'i = ProjectOnSurface(RInt, Fi)
20:                     F'j = ProjectOnSurface(RInt, Fj)
21:                     RDiff1 = RegionDifference(projection1, projecttion2)
22:                     RDiff2 = RegionDifference(projection2, projection1)
23:                     if RDiff1 && RDiff2 then                 ▷ F'i ∩ F'j ≠ ∅
24:                         U = F'i, L = F'j
25:                         Add the pair [i, j, {U,L}] in B[i]
26:                     else if (RDiff1 && !RDiff2) || ( !RDiff1 && RDiff2)
    then
27:                         if !RDiff1 && RDiff2 then            ▷ F'i ⊂ F'j
28:                             U = Fi, L = Fj
29:                             Add the pair [i, j, {U,L}] in B[i]
30:                             break
31:                         end if
32:                     else                                     ▷ F'j ⊂ F'i
33:                         U = F'i, L = Fj
34:                         Add the pair [i, j, {U,L}] in B[i]
35:                     end if
36:                 else                                         ▷ F'i = F'j
```

```
37:                 U = F_i, L = F_j
38:                 Add the pair [i, j, {U, L}] in B[i]
39:                 break
40:             end if
41:         end if
42:     end for
43: end for
44: return B
45: end procedure
```

The functions used in the algorithm are the following:

ProjectAlong(obj): project object *obj* on the XY plane along the *Z-direction*.

RegionIntersection(obj1,obj2): calculate the intersection between *obj1* and *obj2*, which are two coplanar closed surfaces.

RegionDifference(obj1,obj2): calculate the difference between *obj1* and *obj2*, which are two coplanar closed surfaces.

ProjectOnSurface(obj1,obj2): calculate the area in *obj2* by projecting *obj1* on the *obj2* along the *Z-direction*, in which *obj1* and *obj2* are two closed surfaces.

The Algorithm 2 shows the process of reconstructing and trimming the spaces. The function *CreateBooleanDifference (obj1, obj2, tolerance)* that is used computes the Boolean difference between *obj1* and *obj2*. *obj1* and *obj2* are assumed to be two closed volumes.

Algorithm 2 The process of reconstructing and trimming sI-spaces based on their tops and bottoms.

Input: F, B, F_n ▹ F is a list of sorted surfaces based on their average Z coordinate; B is the boundary combinations set; F_n is each surface in F.

Output: $sISpaces$ ▹ sI-spaces.

```
 1: procedure sI-SPACE RECONSTRUCTION.(F, B, F_n)
 2:     Space = []
 3:     for each b_i in B do
 4:         for each temp in b_i do
 5:             r_i = CreateSpace(temp{U, L})
 6:             Add the pair [B[0], B[1], r_i] to Space
 7:         end for
 8:         for each i ∈ [0, Size of S] do
 9:             temp = []
10:             for each k ∈ [0, Size of Space] do
11:                 if Result[k][0] == i then
12:                     Add Space[k][2] in temp
13:                 end if
14:             end for
15:             if Size of temp == 1 then
16:                 Add temp[0] in sISpaces
17:             else
```

```
18:             flag = Size of temp − 1
19:             for eachj ∈ [0, flag] do
20:                 Add temp[j] in sISpaces
21:                 obj1 = temp[flag]
22:                 obj2 = temp[j]
23:                 middleSpace = CreateBooleanDifference(obj1, obj2, toleran
24:                 Add middleSpace in sISpaces
25:             end for
26:         end if
27:     end for
28:   end for
29:   return sISpaces
30: end procedure
```

6.2 SEMI-OUTDOOR & OUTDOOR RECONSTRUCTION

This section presents the approach of reconstructing sO-spaces and O-spaces, which is the process to enclose semi-outdoor and outdoor into 3D enclosed volumes and meanwhile keep the topological consistency between the terrain and them. This approach uses footprints, a contact height, and DTM as inputs and the whole procedure consists of three steps as follows (Figure 6.4):

- Step 1: Extract object footprints;
- Step 2: Classify semi-outdoor and outdoor;
- Step 3: Reconstruct 3D spaces.

6.2.1 EXTRACT OBJECT FOOTPRINTS

This research only concentrates on roads and green areas, because they are the main semi-outdoor and outdoor environments, where people have activities. Moreover, they are generally open or surrounded by objects like the fence, hand railing, enclosing wall, or building walls. For instance, a playground or yard is separated by a fence from the street; a disabled passageway is surrounded by hand railings; an alley is formed by buildings on both sides. To put it another way, areas (roads and green) and objects (fence, hand railing, enclosing wall, or building walls) are two indispensable factors in the formation of sO-spaces. Therefore, in this step, footprints of roads, green areas and objects (fences, hand railings, enclosing walls, and building walls) are extracted from a 2D map.

Furthermore, the I-spaces and sI-spaces are generally originating from volumetric objects (buildings). Thereupon, this research only considers extracting footprints of the I-spaces and sI-spaces formed by built structures. Figure 6.5 illustrates three cases of I-spaces and sI-spaces, in which footprints of indoor and semi-indoor space are extracted based on buildings.

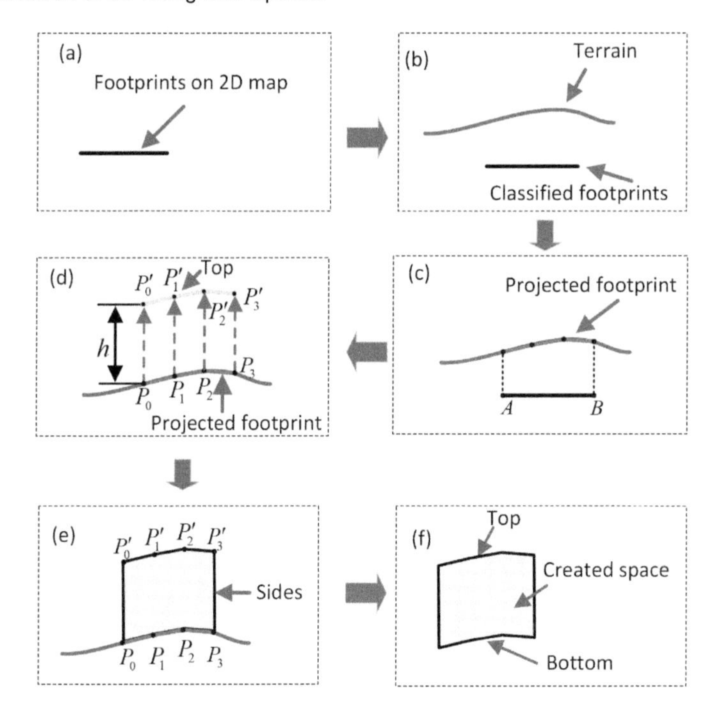

Figure 6.4 The process of sO-space and O-space reconstruction.

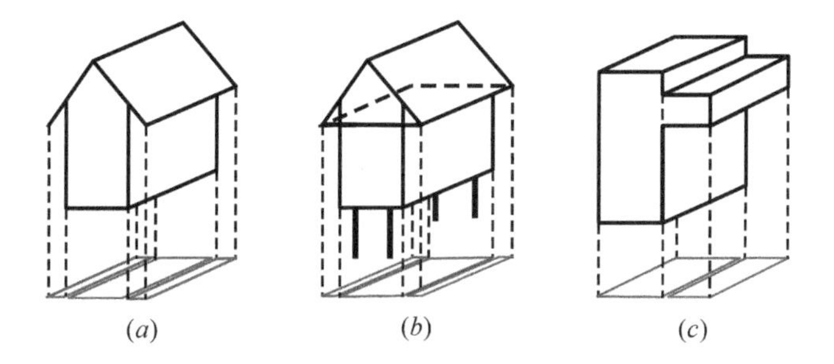

Figure 6.5 The footprints of indoor (red) and semi-indoor (blue) spaces formed by built structures. (a) a building with eaves; (b) a building with eaves and its lowest floor is above ground; (c) a building with a hanging part.

6.2.2 CLASSIFY SEMI-OUTDOOR AND OUTDOOR

The procedure of classifying semi-outdoor and outdoor is based on side closure (C^S). In particular, for each footprint (polygon) of a road or green area, it will be used to have intersection operations with all the surrounding physical boundaries (curves). If the result of intersection operations is a polyline, rather than a point or nothing, it means this polygon is bounded by the physical boundaries (at least partly). Then, the

C^S of each area (polygon) can be computed by Equation 1.5. After that, all footprints are classified into semi-outdoor and outdoor on the basis of the side closure.

6.2.3 RECONSTRUCT 3D SPACES

In this step, the original footprints firstly are projected onto the terrain to get the projected footprints. After that, virtual side boundaries are computed by extruding them up along the *Z-direction* based on the height (h) (Figure 6.4(d)). Here, setting the h is a tricky issue, because there are generally no physical structures that can be used as the references for determining the height of such kind of spaces. In this research, it is set as a contact height (2 meters), because we consider this height is sufficient for pedestrian navigation (Figure 1.6). To make sure the final 3D spaces are correct volumes, the orientation of each polygon is set to counter-clockwise. In particular, suppose there is a segment of projected footprint, which has four vertices $P_0(x_0,y_0,z_0)$, $P_1(x_1,y_1,z_1)$, $P_2(x_2,y_2,z_2)$, and $P_3(x_3,y_3,z_3)$. Then, after counter-clockwise sorting, all the vertices will be extruded up to get $P'_0(x_0,y_0,z_0+h)$, $P'_1(x_1,y_1,z_1+h)$, $P'_2(x_2,y_2,z_2+h)$, and $P'_3(x_3,y_3,z_3+h)$. Then, these vertices are connected end to end to form a side boundary, which can be represented as a $polygon(P_0,P_1,P_2,P_3,P'_3,P'_2,P'_1,P'_0,P_0)$ (Figure 6.4(e)).

Then, 3D spaces (volumes) are reconstructed by joining top(s), sides and bottom(s). The bottom surface is computed by making polygon surfaces from all vertices of the projected footprint. The reconstruction of top is similar, where the vertices are P'_0, P'_1, P'_2, and P'_3. Reconstructed 3D spaces are shown in Figure 6.4(f), where the type of reconstructed space is the same as the corresponding classified footprint, for instance, if a space is reconstructed from a semi-outdoor footprint, the reconstructed space is a sO-space.

6.2.4 ILLUSTRATION

In order to simulate a scene that has indoor, semi-indoor, semi-outdoor and outdoor environments at the same time, this illustration is shown based on a synthetic data, which has a building with eaves, a garden with fence, a pavement, and a street (Figure 6.6(a)). The footprints of these objects subdivide the 2D map of this area as several polygons (Figure 6.6(b)). The polygons (footprints), polygon(A,B,C,D,E,F,A) is the building, polygon(B,C,D,E,K,L,B) is the sI-space, polygon(E,F,G,H,I,J,K,E) is the garden, polygon(L,K,J,I,H,N,M,L) is the pavement, and polygon(M,N,O,P,M) is the street. polyline(J,K,E,F,G,H,I) and polyline(C,B,A,F,E,D) represent physical fence and building walls, respectively, which are shown in double black and red lines.

Except for the footprints of indoor (in red) and semi-indoor (in blue), there are four footprints, fence, garden, pavement, and street (Figure 6.7). Using fence and building walls as physical boundaries, the C^S of the four footprints can be estimated. For the garden, its $l\_e = l_{JK} + l_{KF} + l_{FG} + l_{GH} + l_{HI} = 9 + 10 + 9 + 10 - 2 = 36$, while $l\_t = l\_e + L_{IJ} = 36 + 2 = 38$. Thus, $C^S = l\_e/l\_t = 36/38 = 0.95$. For the pavement, $l\_e = l_{HK} - l_{IJ} = 10 - 2 = 8$, and $l\_t = l_{HI} + l_{IJ} + l_{JK} + l_{KL} + l_{LM} + l_{MN} + l_{NH} = 22 +$

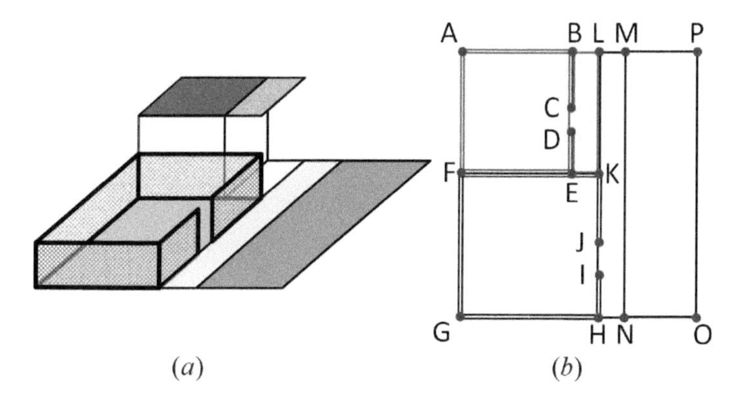

Figure 6.6 Built objects and their non-overlap footprints. (a) Built objects; (b) Corresponding 2D footprints.

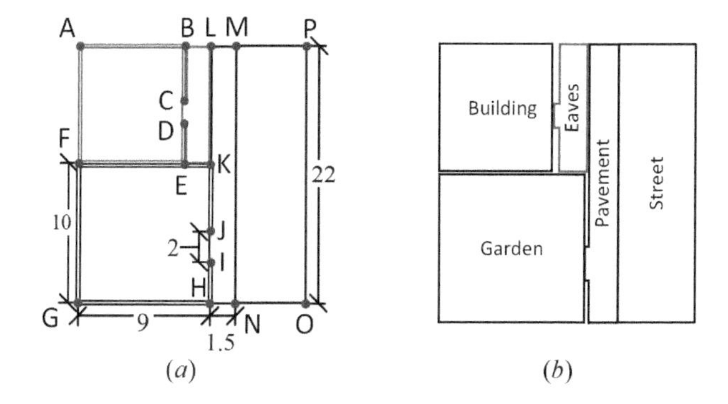

Figure 6.7 Example of footprints classification based on overlaps (unit: m).

$1.5 + 22 + 1.5 = 47$, so the $C^S = l\_e/l\_t = 8/47 = 0.17$. There is no physical boundary for the street. Therefore, the C^S of the garden footprint is 0.95, pavement is 0.17, and the street is 0. Based on the condition that if $C^S \geq 0.75$, the garden is classified as sO-space, while the pavement and street are outdoor spaces (Figure 6.7(b)).

The reconstructed sO-space (garden) and two O-spaces (pavement and street) are illustrated in Figure 6.8. Pedestrians can circulate between the garden space and pavement space, because the two spaces are sharing a virtual boundary surface, so do the pavement and street spaces. But pedestrians cannot directly travel from the garden to street space as the two spaces are isolated from each other.

6.2.5 ALGORITHMS

The complete the process of footprints classification, space reconstruction, and space classification, are described in the provided pseudo code in Algorithm 3, 4, 5 respectively:

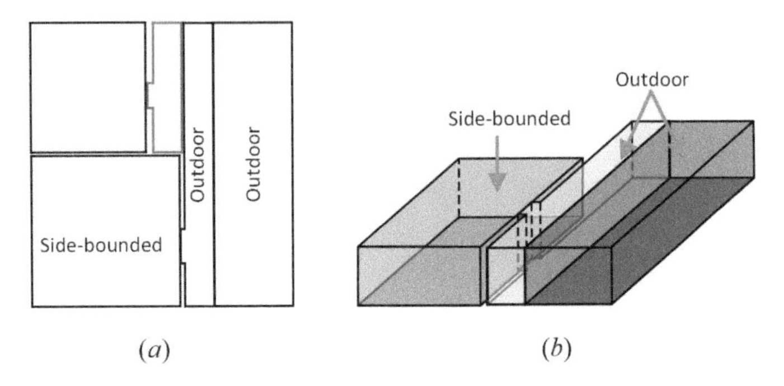

Figure 6.8 Example of reconstructed sO-spaces and O-spaces (height = 5). (a) Classified footprints; (b) reconstructed sO-spaces and O-spaces.

Algorithm 3 Footprints classification based on if $C^S \geq 0.75$.

Input: FP, PB ▹ The *FP* and *PB* are lists including footprints of roads, green areas and physical side boundaries, respectively.

Output: $SBindex, Oindex$ ▹ Two lists include all indices of semi-outdoor and outdoor spaces respectively.

```
1:  procedure FOOTPRINT CLASSIFICATION(FP, PB)
2:      SBindex = [], Oindex = []
3:      N ← size(FP)
4:      for i ← 1 to N do
5:          interL ← 0
6:          fpL ← Length(FP[i])                          ▹ FP[i] is a polyLine
7:          M ← size(PB)
8:          for j ← 1 to M do
9:              inter ← Intersection(PB[j], FP[i])
10:             if isPolyLine(inter) then
11:                 interL ← interL + Length(inter)
12:             end if
13:         end for
14:         C^S ← interL/fpL
15:         if C^S ≥ C^S_0 then
16:             append i to SBindex
17:         else
18:             append i to Oindex
19:         end if
20:     end for
21:     return SBindex, Oindex
22: end procedure
```

Algorithm 4 Space reconstruction based on footprints.

Input: $FP, H, terrain$ ▹ The FP is the same as above; H is a list including heights of all spaces; $terrain$ is a mesh surface.

Output: $Spaces$ ▹ A list including all created spaces.

1: **procedure** SPACE CREATION(FP)
2: $N \leftarrow size(FP)$
3: **for** $i \leftarrow 1$ to N **do**
4: $V \leftarrow getVertices(FP[i])$
5: $V \leftarrow makeCounterClockwise(V)$
6: $topV = [], bottomV = [], sides = []$
7: $M \leftarrow size(V)$
8: **for** $j \leftarrow 1$ to M **do**
9: $edge \leftarrow Line(V[j], V[j+1])$
10: $projectedEdge \leftarrow projectTo(edge, terrain)$
11: $projectedV \leftarrow projectedEdge.Vertices$
12: $projectedV' = [], oneSideV = []$
13: **for** $each\ v\ in\ projectedV$ **do**
14: $v' \leftarrow Point3D(v.X, v.Y, v.Z + H[i])$
15: $append\ v\ to\ oneSideV$
16: $append\ v\ to\ bottomV$
17: $append\ v'\ to\ projectedV'$
18: $append\ v'\ to\ topV$
19: **end for**
20: **end for**
21: $top \leftarrow makePolygonSurf(topV)$
22: $bottom \leftarrow makePolygonSurf(bottomV)$
23: $boundarySurf \leftarrow top + sides + bottom$
24: $oneSpace \leftarrow joinBoundary(boundarySurf, 0.01)$
25: $append\ oneSpace\ to\ spaces$
26: **end for**
27: **return** $Spaces$
28: **end procedure**

Algorithm 5 Space classification based on index.

Input: $Spaces, SBindex, Oindex$ ▹ $Spaces$ is the output of Algorithm 2 and $SBindex$, $Oindex$ are outputs of the Algorithm 1.

Output: $SBSpace, OSpace$ ▹ Two lists including sO-spaces and O-spaces respectively.

1: **procedure** SPACE CLASSIFICATION($Spaces, SBindex, Oindex$)
2: $N \leftarrow size(SBindex)$
3: **for** $i \leftarrow 1$ to N **do**
4: $append\ Spaces[i]\ to\ SBSpace$

```
5:      end for
6:      M ← size(Oindex)
7:      for j ← 1 to M do
8:          append Spaces[j] to OSpace
9:      end for
10:     return SBSpace, OSpace
11: end procedure
```

The functions used in the algorithms are following:

Intersection*(curve, polygon)*: compute the intersection between *curve* and *polygon*. The result can be a polyline, a point, or None.

isPolyLine*(obj)*: judge if the *obj* is a polyline.

Length*(polyLine)*: compute the length of a *polyLine*.

getVertices(obj): get all vertices from a polyline object (*obj*).

makeClockwise*(V)*: make a set of 3D points are ordered by clockwise, where the input parameter *V* is a vertices list.

makeCounterClockwise*(V)*: make a set of 3D points ordered by counter-clockwise, where *V* is a vertices list.

projectTo*(obj1, obj2)*: project *obj1* onto a *obj2* along the *Z-direction*, in which *obj1* is a 3D point list and *obj2* is a mesh surface.

Line*(p1,p2)*: Create a new line segment with two 3D points. *p1* and *p2* will be used as start and end point.

makePolygon*(l_List)*: taking a group of ordered line segments as boundaries to make a polygon.

makePolygonSurf*(v_List)*: taking a group of ordered 3D points as vertices to make a surface.

joinBoundary*(boundarySurface, tolerance)*: Join boundary surfaces in the input array (*boundarySurface*) at any overlapping edges with the *tolerance* to form enclosed volume.

6.3 BUILDING SHELLS RECONSTRUCTION

The inputs for reconstruction of building shells in this research are building footprints, building heights, and DTM. These data are generally available from governmental and open datasets (e.g., OSM). After obtaining building footprints and DTM, the first step is checking if the data have the same coordinate reference system (CRS). If the data are obtained from different sources, they need to be transformed (aligned) in the same CRS. Then, the building shells reconstruction process consists of four steps as follows (Figure 6.9):

- Step 1: Compute TIC by projecting footprints onto the terrain;
- Step 2: Set height and create sides;
- Step 3: Generate top and bottom to build building shells;
- Step 4: Rebuild terrain considering TIC as constraints.

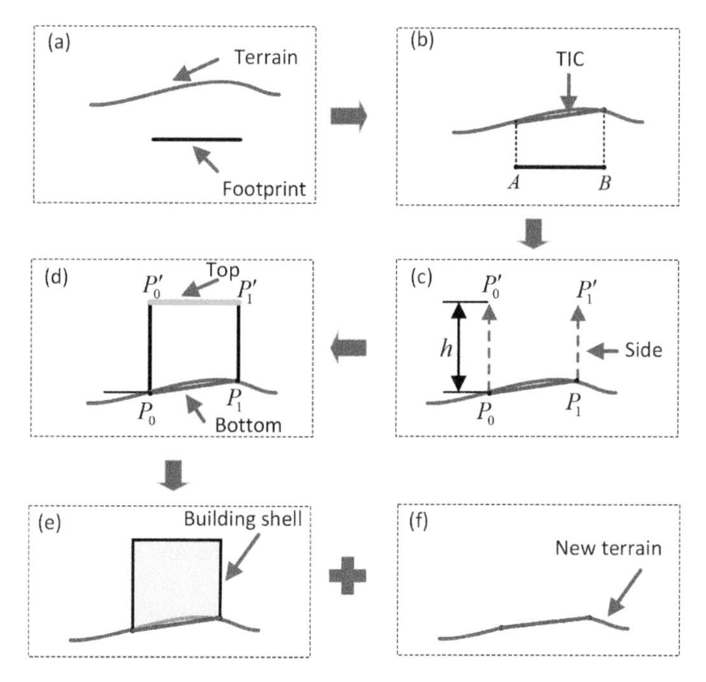

Figure 6.9 The process of building shells reconstruction based on building footprints, building heights and DTM.

6.3.1 COMPUTE TIC BY PROJECTING FOOTPRINTS ONTO THE TERRAIN

The footprints are normally presented as 2D polygons with coplanar vertices. Therefore, the first step is to compute the corresponding building footprints projection on the terrain. This projections of building footprint on the terrain are the TIC.

The TIC computation process is **P**rojecting **V**ertices of footprints on the **T**errain (PVT). This practice ensures the number of points of the original footprints is preserved (Figure 6.9(b)). As expected, the shape of the projected footprint is unaltered as indicated from the top view, whereas the terrain is changed.

6.3.2 SET HEIGHT AND CREATE SIDES

After getting each counter-clockwise oriented projected footprint, sides are reconstructed by extruding them up along the *Z-direction* based on building heights (h) (Figure 6.9(c)). In particular, a line segment of a projected footprint, which has the start and end vertices $P_0(x_0, y_0, z_0)$ and $P_1(x_1, y_1, z_1)$, respectively. P_0 and P_1 will be extruded up along the *Z-direction* based on h to get $P'_0(x_0, y_0, h)$ and $P'_1(x_1, y_1, h)$. Then, these four vertices can form a side, which can be represented as a $polygon(P_0, P_1, P'_1, P'_0, P_0)$. It is worth noting that each composing polygon should follow counter-clockwise orientation to make sure the final 3D spaces are correct volumes (Figure 6.10).

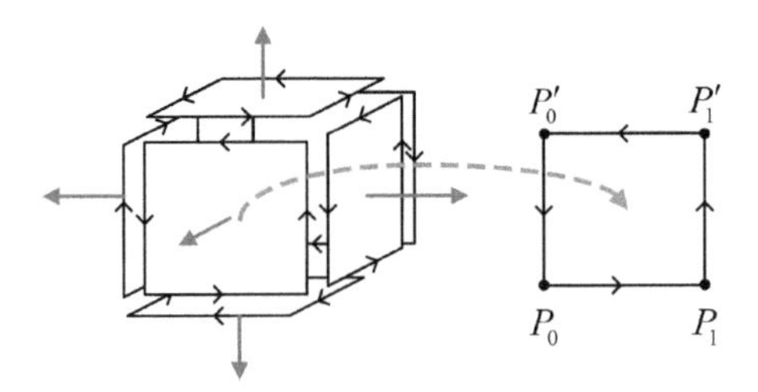

Figure 6.10 The outward-facing normal of boundaries of a 3D building shells.

6.3.3 GENERATE TOP AND BOTTOM TO RECONSTRUCT BUILDING SHELLS

The third step is to generate the top(s) and bottom(s), and then join them with the reconstructed sides to make 3D volumes (Figure 6.9(d) and (e)). All tops are considered as planar surfaces, and they are generated by making a polygon surface from a set of vertices. The bottom is slightly different from the top, because vertices used for the bottom(s) are not coplanar. Thus, after connecting the projected vertices , we can get an enclosed 3D polyline, and then it is patched as a bottom surface. After getting the top surfaces, bottom surfaces, and side surfaces, each 3D building shell can be reconstructed by joining them.

6.3.4 REBUILD TERRAIN CONSIDERING TIC AS CONSTRAINTS

The process of rebuilding terrain is taking the vertices of projected footprints as Points, and their edges as Breaklines (constraints) to calculate a Constrained Delaunay Triangulation (CDT), see Figure 6.9(f).

6.3.5 ILLUSTRATION

To illustrate and evaluate the entire process of the approach, a precinct example is employed, which includes 98 building footprints (Figure 6.11). The original building footprints are provided as a CAD file, in which all the footprints are polygons. The height of each buildings are also prepared. A DEM (5 metre) of the example area is used to construct Delaunay triangulation representing the terrain (DTM).

As the footprints and DTM are in the same coordinate system, they aligned automatically after importing. Then, the whole process starts from projecting the building footprints onto the terrain (Figure 6.12). From the top view, projected footprints (TIC) have the same shape as the original footprints, but they are 3D polylines rather than polygons.

After aligning the building footprints with terrain based on their geolocation, the building footprints are projected onto the terrain. As mentioned above, there are two methods to get the projected footprints (Figure 6.12), PVT and PPT. From the top view, the shape of projected footprints is the same as the original, but the results of PVT are 3D polylines, while the PPT method are mesh surfaces.

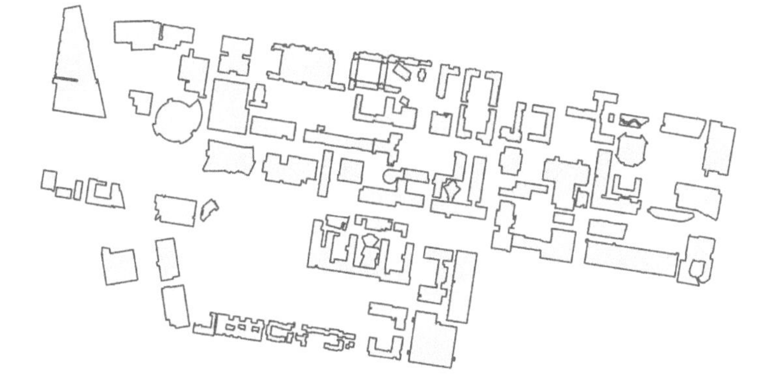

Figure 6.11 Building footprints for illustration.

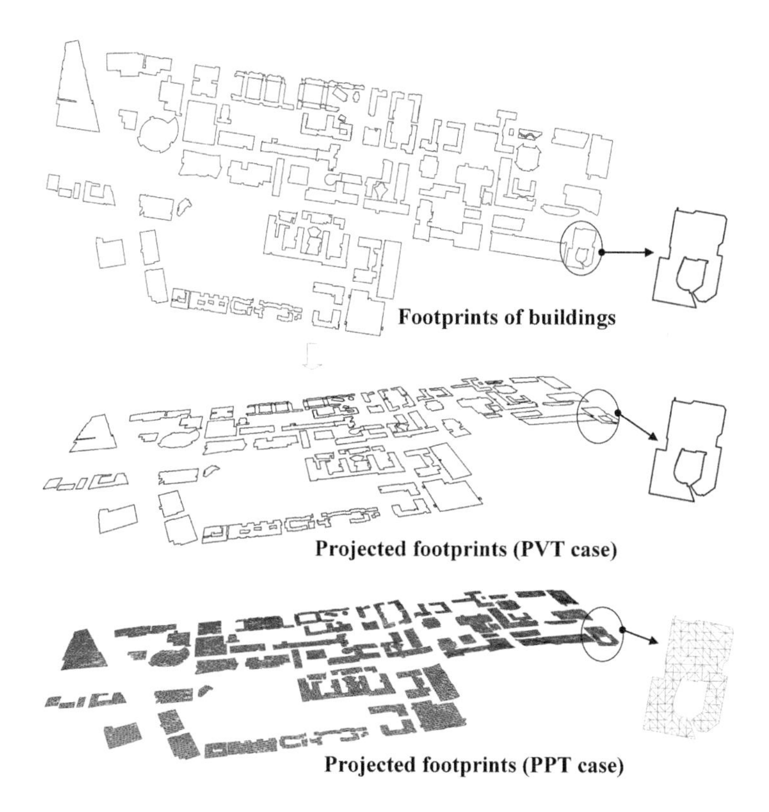

Figure 6.12 The projected footprints on the terrain.

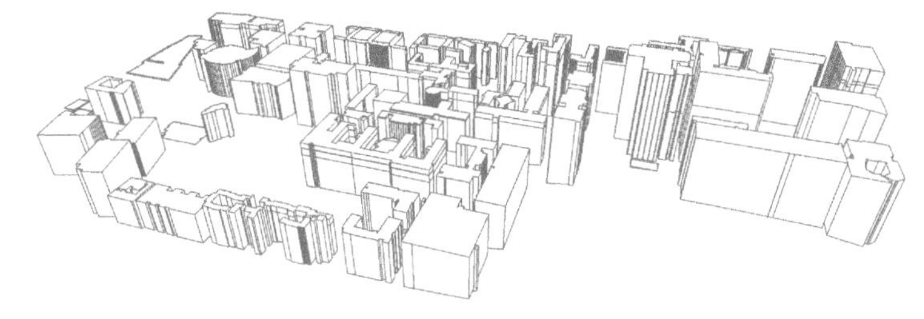

(a) The reconstructed building shells.

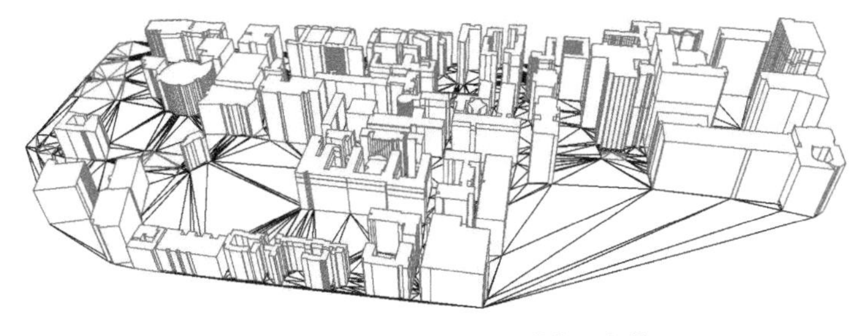

(b) The terrain combined with building shells.

Figure 6.13 The building shells and the terrain that are reconstructed based on PVT method.

Before the walls generation procedure, 3D polylines (TIC) of the projected footprints are extracted. For the PVT method, the projected footprints are already captured as 3D polylines, but for PPT method, the edges with a single adjacent face will be extracted. Then, each line segment of a polyline will be extruded up to the same height along the *Z-direction* based on building heights to get a quadrilateral side walls. The bottom is made by patching the 3D polyline as a surface. The top is a planar surface constructed by ordered 3D points. Then, the building shells are enclosed by joining the tops(s), sides, and bottom(s) (Figure 6.13(a)). The component named *Fragment Patch* in *Grasshopper* is used, which takes *Fragment polyline boundary (curve)* as inputs, and *Fragmented patch* as outputs. The roof is a planar surface constructed by ordered 3D points.

The final step is to re-compute terrain using the vertices and edges of the projected building footprints as constraints. The results of the re-triangulated terrain show that footprints are clearly recognizable within the triangulated surface. The principle of this method is changing the 3D I-spaces and terrain slightly to fit each other. The results show that the topological issue has been fixed and building shells are reconstructed successfully (Figure 6.13(b)).

Another method is **P**rojecting the whole footprint **P**olyline on the **T**errain (PPT) (Figure 6.14). This method ensures the curvature of the terrain but might lead to unnecessarily complex walls, which differ from the reality (Figure 6.14(a)), but it has a more detailed terrain (Figure 6.14(b)). Comparing terrain of the two methods,

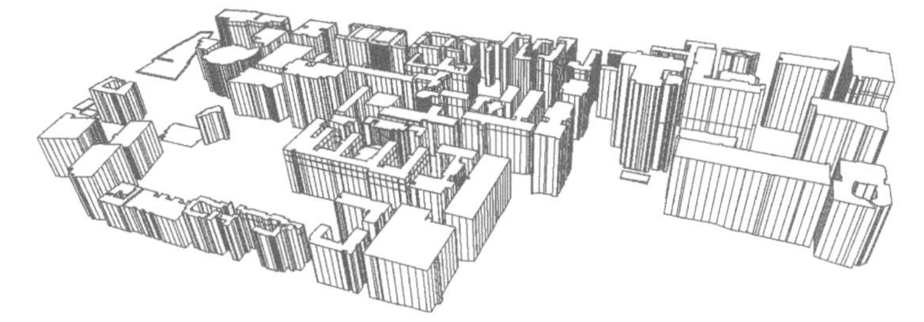

(a) The reconstructed building shells.

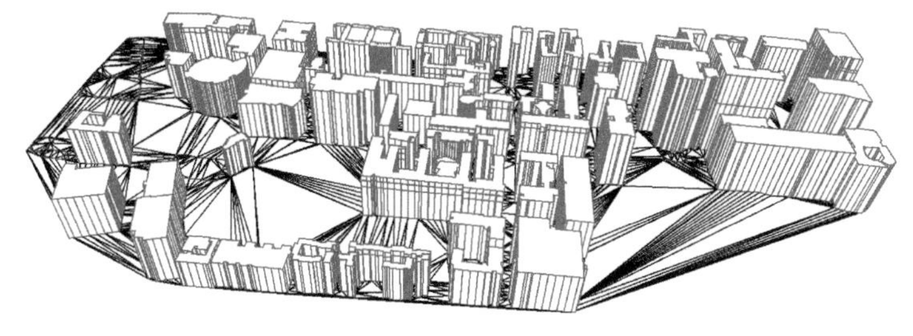

(b) The terrain combined with building shells.

Figure 6.14 The building shells and the terrain that are reconstructed based on PPT method.

the result of PVT has less vertices and triangles in the terrain (Figure 6.13(b)), which means that the changes of terrain are more than in the PPT method.

The results of the two re-built DTM show that footprints are clearly recognizable within the triangulated surface. That is, the topological consistency issue between the reconstructed 3D building shells and the terrain is successfully solved. The main point of the two methods are the reconstruction of building shells. The PVT method is more efficient in one aspect, having less vertices and facades, which is beneficial for a large precinct scale area. The PPT method complicates the building shells, but it preserves more details of TIC, which is good for estimating the areas of the building facades. (Figure 6.14(b)). It should be clarified that the terrain from PPT method has more triangles than the PVT method, but we cannot conclude that the terrain of the former is more precise than that of the latter method. Because the only difference comes from the TIC and only if there are more points in the empty spaces for terrain reconstruction, the rebuilt terrain will be more precise.

This illustration used the function of the transformer (*TINGenerator*) from the software FME (Feature Manipulation Engine), where the vertices of projected footprints are used as *Points/Lines*, and their edges are *Breaklines*.

6.3.6 ALGORITHM

The whole process, except the terrain re-computation, can be followed in the provided pseudo code in Algorithm 6:

Algorithm 6 The process of reconstructing building shells.

Input: $FootPrint, h$ ▷ The *FootPrint* and *h* are lists including footprints of space and heights respectively.

Output: $Buildings$ ▷ 3D enclosed building objects.

1: **procedure** BUILDING SHELLS RECONSTRUCTION.($FootPrint, h$)
2: $Buildings = []$
3: **for** *each footprint in FootPrint* **do**
4: $V = getVertices(footprint)$
5: $V = makeCounterClockwise(V)$
6: $V' = projectTo(V, terrain)$
7: $V'' = (V_X,\ V_Y,\ footprint.h)$
8: $Sides = []$
9: **for** *each i in*$[0, Size\ of\ V]$ **do**
10: $l1 = Line(V_i, V_{i+1})$
11: $l2 = Line(V_{i+1}, V''_{i+1})$
12: $l3 = Line(V''_{i+1}, V''_i)$
13: $l4 = Line(V''_i, V_i)$
14: $l\_List = [l1, l2, l3, l4]$
15: $sidePolygon = makePolygon(l\_List)$
16: *Add sidePolygon to Sides*
17: **end for**
18: $Top = makePolygonSurf(V'')$
19: $V' = makeClockwise(V')$
20: $Bottom = makePolygonSurf(V')$
21: $boundarySurf = [Top, Sides, Bottom]$
22: $building = joinBoundary(boundarySurf, 0.01)$
23: *Add building to Buildings*
24: **end for**
25: **return** *Buildings*
26: **end procedure**

The functions used in the algorithm are the following:

getVertices(obj): get all vertices from the *obj*, where the *obj* is a polyline.

makeClockwise(V): make a set of 3D points ordered in clockwise direction, where the input parameter *V* is a list of vertices.

makeCounterClockwise(V): make a set of 3D points ordered in counter-clockwise direction, where *V* is a list of vertices.

projectTo(obj1, obj2): projecting *obj1* on a *obj2* along the *Z-direction*, in which *obj1* is a 3D point list, and *obj2* is a mesh surface.

Line(p1,p2): Constructs a new line segment between two 3D points. $p1$ and $p2$ will be used as start and end point.

makePolygon(l_List): taking a group of ordered line segments as boundaries to make a polygon. The parameter, *l_List*, is a line list.

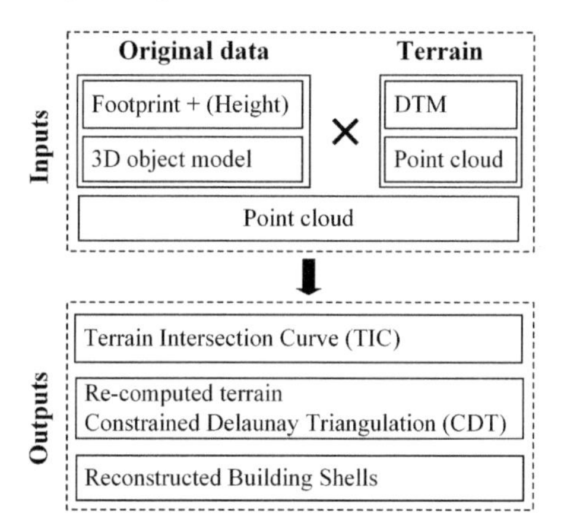

Figure 6.15 The other data sources for building shells reconstruction.

makePolygonSurf(v_List): taking a group of ordered 3D points as vertices to make a surface, where *v_List* is a list of vertices.

joinBoundary(boundarySurface, tolerance): Joins the boundary surfaces in the input array (*boundarySurface*) at any overlapping edges with the *tolerance* to form enclosed volume.

6.3.7 OTHER POSSIBLE APPROACHES OF BUILDING SHELLS RECONSTRUCTION

In addition to the presented approach, other possible approaches of building shells reconstruction can be distinguished based on data source (Figure 6.15): (i) footprints + Point cloud, (ii) 3D objects + DTM, (iii) 3D objects + 3D point clouds, and (iv) Point cloud only.

6.3.7.1 Footprints + Point Cloud

With the wide application of laser scanning technology, 3D point clouds become a data source for 3D city objects [53]. Hence, footprints and point clouds are also a possible input combination to construct 3D models and terrain. The proposed approach includes four steps (Figure 6.16):

- Step 1: Classify point cloud based on footprints; This approach assumes the original point clouds are unclassified. Footprints are afterwards used to group point clouds for each building. The terrain can be also derived from the point clouds [207].
- Step 2: Compute roof outline; After grouping the point clouds, the average *Z* of roofs will be calculated. Then, using the *Z* of roof points and the footprint, we can get the outline of a roof. It should be mentioned that this step actually simplifies the roof as a horizontal planar, rather than pitched.

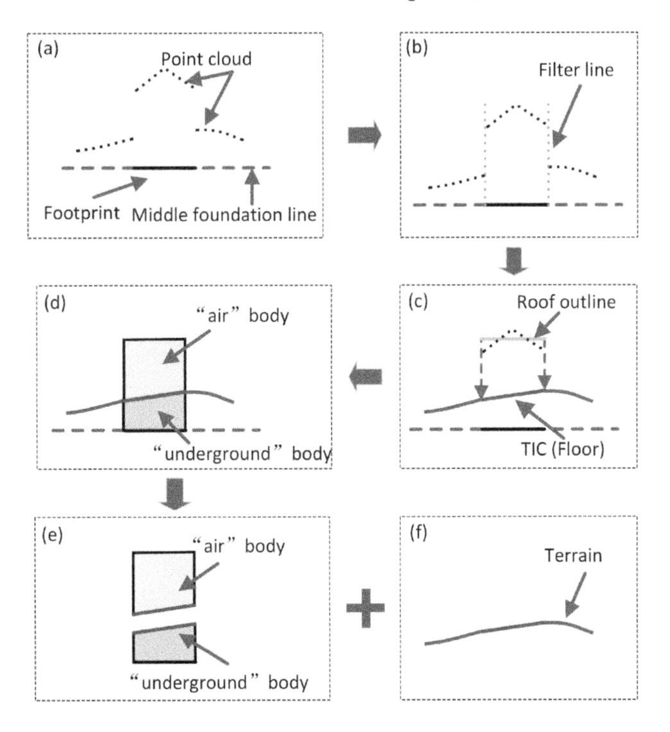

Figure 6.16 The process of modeling based on footprint and point cloud.

Surely, if the roofs are represented as tilted polygons matching the point cloud distribution, the final results of 3D buildings will be more accurate.

- Step 3: Project roof outline onto terrain to get TIC and 3D building (solid); This step is projecting the roof outline onto the terrain along the *Z-direction* to get the TIC. We can get the wall surface of "air" body by connecting the planar roof with floor contour (generated based on TIC) (Figure 6.16(c)). The floor contour will be further patched into floor surface, as is the roof. Combining the roof surface, side walls, and floor surface, the "air" body of the building are created. If we have point clouds demarcating the underground, they can be processed with the similar steps to get the "underground" body (Figures 6.16(d) and (e)).
- Step 4: Rebuild terrain; The last step is using the TIC as *Breaklines*, and the terrain point clouds as *Points* to rebuild a CDT (Figure 6.16(f)).

6.3.7.2 3D building model + DTM

This method of creating 3D building model is very popular in current applications for the creation of large-scale 3D models, because it is one of simplest way to generate them as extruded footprints to a certain height [208]. Also, the BIM/CityGML are two popular sources of 3D building models, which has been widely used in different domains, such as urban planning, city analytic [209, 210].

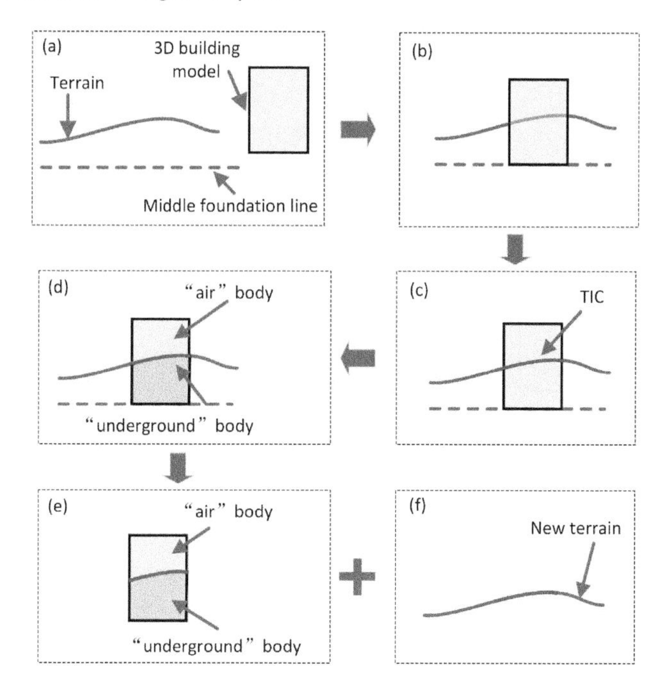

Figure 6.17 The process of modeling based on 3D building models and DTM.

This approach taking 3D building model and DTM as inputs includes four steps (Figure 6.17).

- Step 1: Align 3D building model with DTM; This alignment step is used to bring georeferencing information to 3D building models. In other words, if the 3D building models are georeferenced already or have georeferencing information (e.g., BIM models have a translation vectors), this step is not needed. Otherwise, this step is used to make sure that the 3D building models are at the right location, not only considering horizontal direction but also the vertical (Figure 6.17(b)).
- Step 2: Get TIC by 3D intersection; After the alignment, the 3D building model will have an intersection with terrain geometrically. Thus, the TIC can be computed by 3D intersection between the 3D building models and terrain (Figure 6.17(c)).
- Step 3: Obtain "air" body and "underground" body; The 3D building model are cut into "air" body and "underground" body (if it has) for some applications. For instance, doing area budget for wall decoration, where the walls of "air" body are needed. Then, the TIC will be used to make a surface, and this surface will be used twice to cap the upper body and lower body as enclosed volumes. In other words, the "air" body and "underground" body are volumes, and the floor surface of the former and the roof surface of the

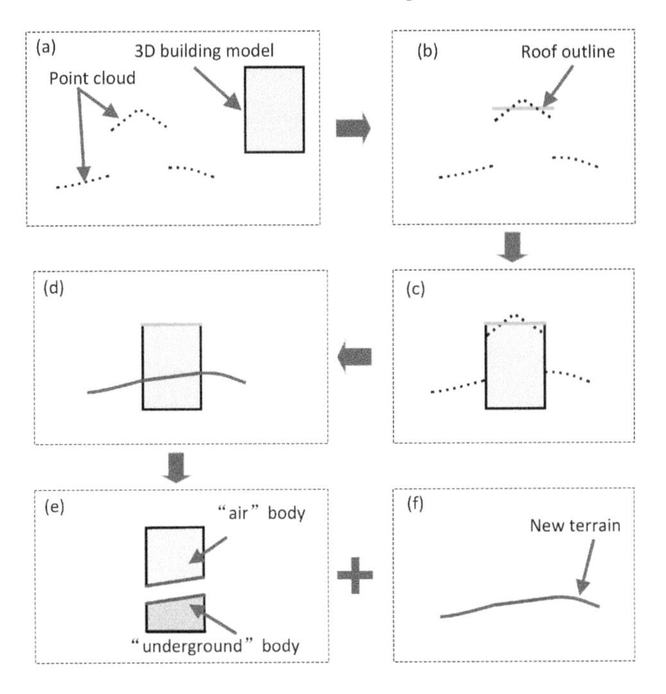

Figure 6.18 The process of modeling based on 3D building models and point clouds.

latter is actually the surface patched from TIC (Figures 6.17(d) and (e)) .

- Step 4: Re-compute terrain; A new terrain will be computed based on the original DEM and TIC (Figure 6.17(f)). The geometry of DEM will be coerced into points or multi-points. Then, a CDT will be generated by taking the points beyond the TIC as *Points*, and TIC as *Breaklines*. This re-computation keeps original shapes of terrain except the part occupied by 3D objects.

6.3.7.3 3D building model + Point Cloud

When taking 3D building models and point clouds as the inputs, we have to keep in mind that at least one of them should be georeferenced and point clouds should be classified into different categories, such as building, terrain, and vegetation. Thus, footprints are needed to ensure that the 3D building models appear in the right locations, and point cloud representing roofs should be classified by footprints, which is similar to the Figure 6.16(b).

With georeferenced 3D building models, and classified point clouds, the proposed approach includes four steps (Figure 6.18):

- Step 1: Estimate outline of a roof face; With the classified roof point clouds, it is possible to estimate the orientation and height of a roof face accurately, but the outline of a roof face is a difficult issue [211]. Therefore, we simplify this step by representing all roofs as planar surfaces. Thus, for each roof, its point cloud will be used to compute a planar roof based on the average *Z* (Figure 6.18(b)).

- Step 2: Align 3D building models with roof outlines; Then, 3D building models will be aligned based on their corresponding roof lines. If the roof of a 3D building is planar, it will be moved to overlap with roof lines. Otherwise, for a pitched roof, the middle line of the roof in the *Z-direction* will be used to align with the roof line.
- Step 3: Construct initial terrain and TIC; For the terrain point cloud part, an initial terrain can be constructed by using triangulation for instance. With these alignments, the 3D building model will have an intersection with the terrain. Thus, the TIC can be computed by 3D intersection between the 3D building model and terrain (Figure 6.18(d)).
- Step 4: Rebuild 3D building models and terrain; The last step is recomputing a 3D building model as 'air' body and 'underground' body (if it has) based on TIC (Figure 6.18(e)). Moreover, a CDT will be re-computed as new terrain by taking the terrain point cloud as *Points*, and TIC as *Breaklines*(Figure 6.18(f)).

6.3.7.4 Point Cloud

The consideration of taking only point clouds as data source for both 3D buildings and terrain assumes that point clouds are georeferenced and classified into at least two categories, buildings and terrain. If not, point clouds should be georeferenced and classified beforehand.

With the georeferenced and classified point cloud, the whole process to construct 3D building models, TIC, and terrain requires four steps (Figure 6.19):

- Step 1: Obtain outline of a roof face; In this approach, we also simplify all roofs as planar surfaces. Thus this step is the same as the first step of the previous approach (*3D building model + Point Cloud*) (Figure 6.19(b)).
- Step 2: Compute original terrain and TIC; An original terrain can be computed by using terrain point cloud. Then, we can project roof outlines onto terrain along the *Z-direction* to get the TIC (Figure 6.19(c)).
- Step 3: Re-compute 3D buildings (solids); Then, walls of "air" body can be obtained based on the roof lines and TIC. The TIC will be further patched into floor surface and roof surface for the "air" body and "underground" body respectively (Figure 6.19(d) and (e)).
- Step 4: Rebuild terrain. The last step is also recomputing a CDT as new terrain by taking the terrain point cloud as *Points*, and TIC as *Breaklines* (Figure 6.19(f)).

6.4 SUMMARY

This chapter shows approaches that can automatically reconstruct sI-spaces, sO-spaces, O-spaces, and building shells as 3D spaces (volume). Thus, all types of

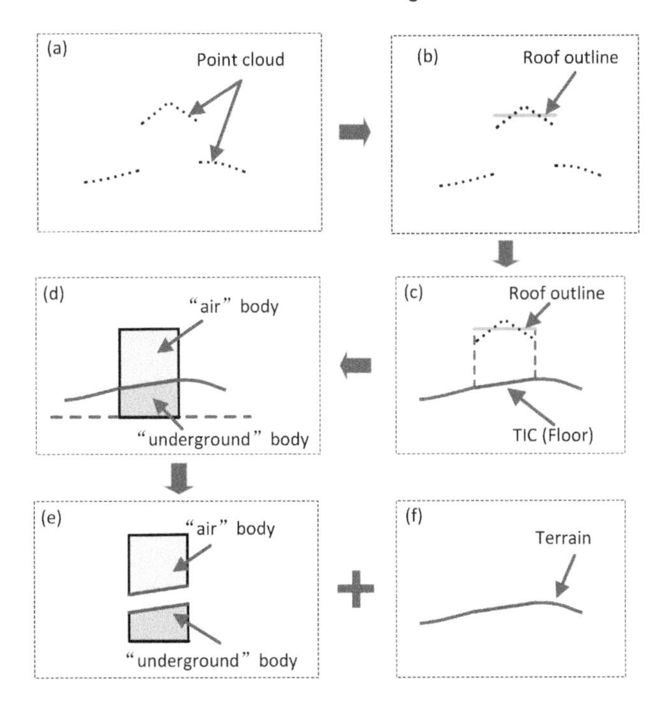

Figure 6.19 The process of modeling based on point cloud only.

spaces are able to mimic the indoor environments to derive a network based on 3D connectivity of spaces.

The sI-space reconstruction is an automatic method based on 3D models. The demonstration indicates that the reconstruction of sI-spaces (formed by built structures) is feasible on the basis of existing 3D standards such as CityGML or BIM. The reconstruction of sO-space and O-space is an automatic method based on a 2D map and DTM. However, considering the biggest limitation of the proposed approach is the data source, i.e., we generally do not have suitable data, such as detailed CityGML LoD 3 datasets or BIM models. The approaches presented are only to demonstrate that these spaces can be created. We believe that the approaches can provide some inspirations for other researchers who need to reconstruct 3D spaces for navigation or other applications. The reconstruction procedure of the building shells is not a novel approach, but considering not always all the buildings exist as indoor models, therefore, the building shells still have to be reconstructed to provide better orientation. This research reconstructs the building shells to represent indoor spaces of the buildings that have no 3D models (e.g., BIM, CityGML LoD 4 models). But it does not mean that the 3D spaces within building shells are equal to indoor spaces as they have no interior.

In short, the sI-spaces formed by the building with 3D models are recommended to be reconstructed based on the projection-based method, while the reconstruction approach of the rests (data source is 2D maps) is extruding footprints of shelters along the *Z-direction* with a contact height (2 meters). The procedures for sO-spaces and O-spaces reconstruction is also based on extrusion. Furthermore, all 3D space reconstructions in this research consider keeping the topological consistency between 3D spaces and terrain based on TIC. It is not true to assume that the reconstructed 3D spaces are standing on a plane surface and this surface is regarded as the default terrain. In fact, the natural terrain is irregular and uneven, rather than planar. It should be noted that the proposed approaches work fine within precinct areas (e.g., a campus or similar), where pedestrians can reach on foot. If the scenarios become a city level, abstraction/aggregation mechanisms will be needed. This research does not reach the complexity and optimization topics, but these topics will be investigated in-depth in future work (Section 8.4.4).

7 Implementation & Case Study

This chapter implements the previously developed theories, models, and approaches for indoor-outdoor seamless navigation path planning. It includes data preparation, space classification and reconstruction, space selection, unified space-based navigation model (navigation network) derivation, path planning, and comparison of results. Finally, the uncertainties and limitations of the whole work are discussed.

7.1 DATA, SOFTWARE, AND FLOWCHART FOR IMPLEMENTATION

The test area for seamless indoor and outdoor path planning, MTC-path, NSI-path is part of the University of New South Wales (UNSW) campus in Kensington (Figure 7.1). There are seven types of objects, including buildings, shelters, roads, green areas, hand railings, enclosing walls and fences (Figure 7.2). The buildings indicate I-spaces, hand railings, enclosing walls and fences serve as physical boundaries. Shelters represents areas of sI-spaces. Roads and green areas are sO-spaces and O-spaces.

The initial data of the objects are footprints in a geo-referenced 2D map. The coordinate reference system (CRS) is EPSG:28356. It is provided by the Estate Management (EM) of the UNSW as a CAD file, in which no attributes are attached to the geometry. General speaking, we can obtain such a 2D map with attributes from OSM, but in this implementation, geometry in the map provided by UNSW EM is used while the map from OSM is only utilized as the source of attributes. There are two reasons: by comparison, the footprints of the buildings in the map provided by EM have more details, and more importantly, this map contains footprints of unknown roads in the selected area.

There are four software packages are employed in this implementation: Quantum GIS (QGIS), Rhinoceros (with Grasshopper), the FME (Feature Manipulation Engine), and PostgreSQL with PostGIS extension. In the whole process, QGIS is used to add attributes, set the coordinate system, and edit the footprints. Rhinoceros (with Grasshopper) is used to process the entire workflow, and the whole data process is developed in *Python* script. FME is used for extracting point clouds from the DEM, and conducting terrain re-computation and visualization. For instance, in the terrain rebuilt (Section 6.3), the transformer named *TINGenerator* from FME is utilized to constructs a Delaunay triangulation based on input points and breaklines. PostgreSQL is the data management tool of original BIM models.

The data preparation includes three steps: (i) manage all the footprints as separate layers in QGIS; (ii) copy the attributes of footprints from OSM to the map provided from EM by utilizing the function named "Join attributes by location" in QGIS; (iii) manually correct all the footprints to be topologically correct. All the footprints of

DOI: 10.1201/9781003281146-7

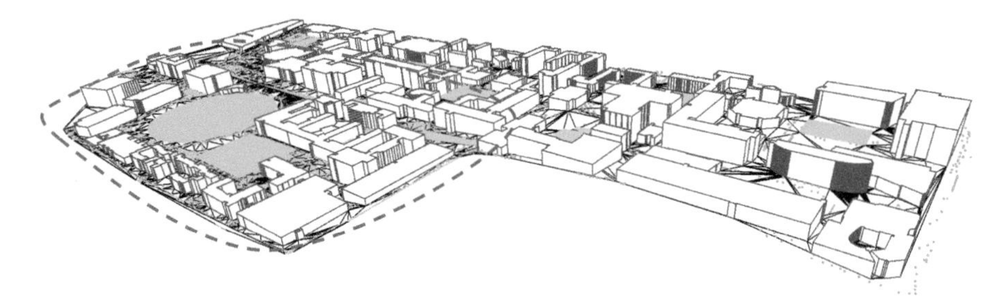

Figure 7.1 Spaces on UNSW campus. The selected part for further implementation is marked by red dash-lines.

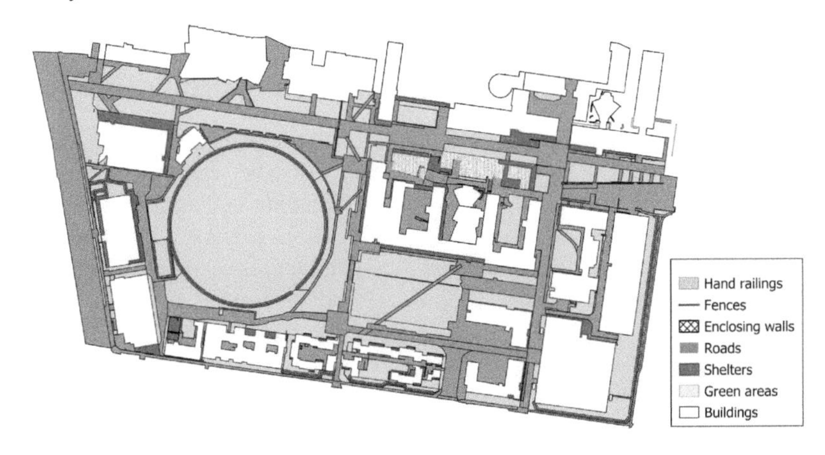

Figure 7.2 Footprints of spaces on the selected area of UNSW campus. The dot-filled footprint is the building that has BIM model.

surface objects are redrawn as closed polygons and they may share edges, but without any overlaps. The roads are connected, especially the footpaths for pedestrians (without names but connected to other roads), so does the green areas. Thus, the continuous roads/green areas are manually subdivided into small polygons (Figure 7.3), in which the following principle is followed: if the area has a name, it will be subdivided from other connected areas. Otherwise, we only consider subdividing the areas based on their attributes, such as if an area is a green area, this area will be redrawn as a closed polygon.

We employ a BIM model of a building (named as Red Center H13) as the source of I-spaces (Figure 7.4(a)). In this model, only rooms and doors are utilized. The original model is stored in PostgreSQL, in which each room/door has seven attributes: $\{ifc\_class, ifc\_guid, ifc\_name, ifc\_description, ifc\_containing\_storey, geom, color\}$. The geometry type of spaces is POLYHEDRALSURFACE Z (Figure 7.4(b)).

Furthermore, a DEM (5 meter) of the test area is obtained from Elevation Foundation Spatial Data [1] that belongs to National Elevation Data Framework (NEDF),

[1] http://elevation.fsdf.org.au/

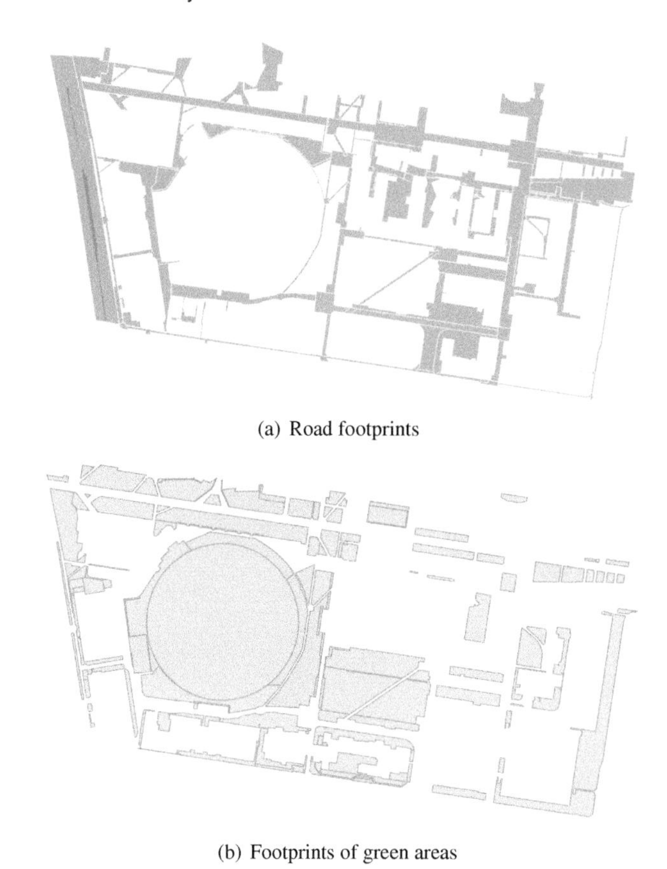

(a) Road footprints

(b) Footprints of green areas

Figure 7.3 Footprints of road and green areas in the selected area of UNSW campus.

Geosciences Australia. The geometry of the DEM is coerced into point clouds, and then, they are used to construct Delaunay triangulation representing the bare terrain (DTM) (Figure 7.5).

The flowchart for implementation is shown in Figure 7.6. The whole process consists of five steps:

(i) Classify road and green spaces as semi-indoor and semi-outdoor by using footprints of building, fence, hand railing, and enclosing wall as physical boundaries;
(ii) Reconstruct sI-spaces, sO-spaces, and O-spaces based on extrusion, in which footprints of shelters are used directly for the reconstruction of sI-spaces;
(iii) Select spaces based on the size (see Section 7.3);
(iv) Derive navigation network based on Poincaré duality (Section 3.1.1);
(v) Plan navigation paths based on Dijkstra algorithm.

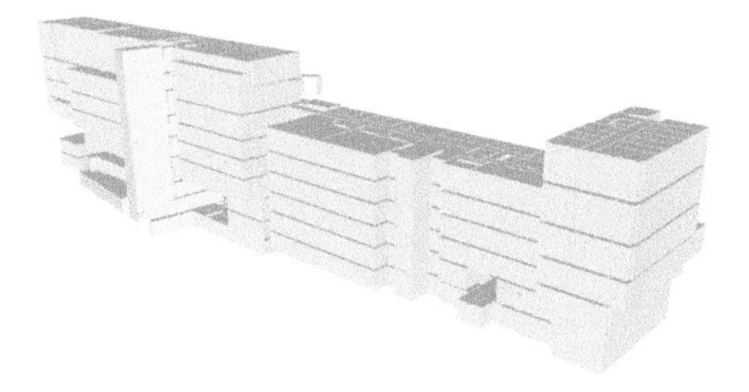

(a) The I-spaces come from the BIM model.

CRC_LCL on postgres@CRC_LCL

```
SELECT ifc_class, ifc_guid, ifc_name,ifc_description,ifc_containing_storey,ST_AsText(geom),color
FROM public."Red_Center_H13_IFC_color"
where ifc_class = 'Spl_IfcDoor' or ifc_class ='IfcSpace'
```

Data Output Explain Messages Query History

	ifc_class character varying	ifc_guid character varying	ifc_name character varying	ifc_description character varying	ifc_containing_st character varying	st_astext text	color character varying
1	IfcSpace	3Kq4JtKQj2SQTB95t...	B03B	Room	Basement	POLYHEDRALSURFACE Z (((16834...	rgb(130, 130, 130)
2	IfcSpace	3Kq4JtKQj2SQTB95t...	BQ03	Room	Basement	POLYHEDRALSURFACE Z (((16834...	rgb(130, 130, 130)
3	IfcSpace	3Kq4JtKQj2SQTB95t...	B03	Room	Basement	POLYHEDRALSURFACE Z (((16834...	rgb(130, 130, 130)
4	IfcSpace	3Kq4JtKQj2SQTB95t...	B05	Room	Basement	POLYHEDRALSURFACE Z (((16834...	rgb(130, 130, 130)
5	IfcSpace	3Kq4JtKQj2SQTB95t...	B01B	Room	Basement	POLYHEDRALSURFACE Z (((16834...	rgb(130, 130, 130)
6	IfcSpace	3Kq4JtKQj2SQTB95t...	B01A	Room	Basement	POLYHEDRALSURFACE Z (((16834...	rgb(130, 130, 130)
7	IfcSpace	3eaNOexAb4BxeyC4...	G022	Room	GroundFloorS...	POLYHEDRALSURFACE Z (((16834...	rgb(130, 130, 130)
8	IfcSpace	3Kq4JtKQj2SQTB95t...	B06	Room	Basement	POLYHEDRALSURFACE Z (((16834...	rgb(130, 130, 130)
9	IfcSpace	3Kq4JtKQj2SQTB95t...	B07	Room	Basement	POLYHEDRALSURFACE Z (((16834...	rgb(130, 130, 130)
10	IfcSpace	3Kq4JtKQj2SQTB95t...	B07A	Room	Basement	POLYHEDRALSURFACE Z (((16834...	rgb(130, 130, 130)
11	IfcSpace	3Kq4JtKQj2SQTB95t...	B08	Room	Basement	POLYHEDRALSURFACE Z (((16834...	rgb(130, 130, 130)

(b) The table of I-spaces in PostgreSQL.

Figure 7.4 The I-spaces and BIM model in PostgreSQL.

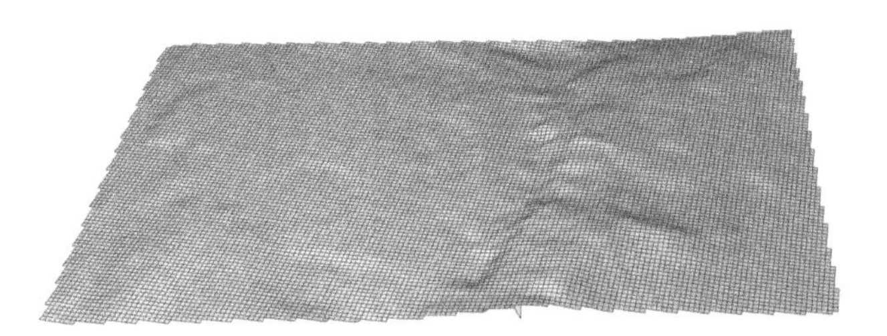

Figure 7.5 The terrain of the precinct.

7.2 SPACE CLASSIFICATION AND RECONSTRUCTION

The conceptual model (Figure 4.1) is partially implemented to demonstrate its use. In addition to the two classes (*IndoorCore::CellSpaceBoundary* and *IndoorCore::CellSpace*) themselves and their attributes, we implement the (association) link between them.

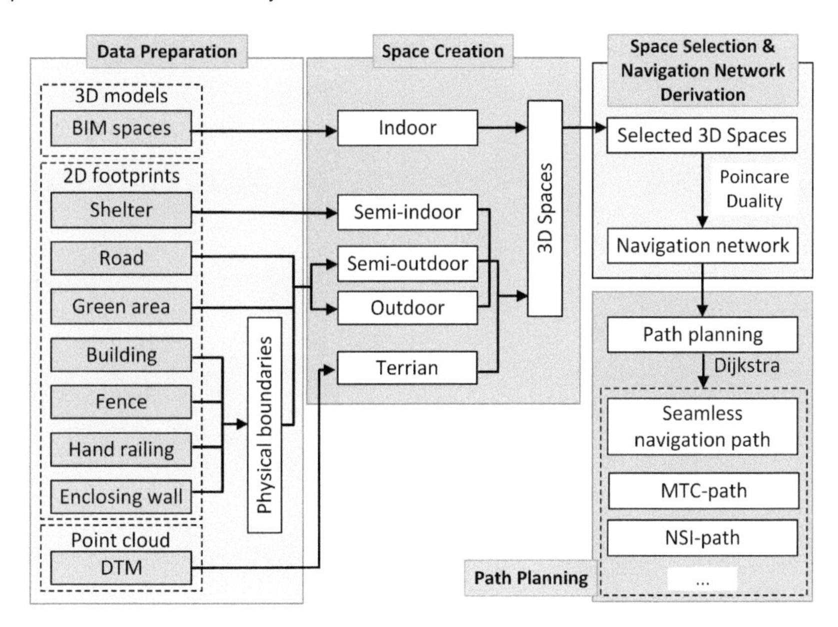

Figure 7.6 Flowchart of the implementation.

```
class CellSpace:
def__init__(self, sType, topClosure, sideClosure, bottomClosure,
containBoundary):
self.sType = 'OSpace'
self.topClosure = topClosure
self.sideClosure = sideClosure
self.bottomClosure = bottomClosure
self.containBoundary = containBoundary
```

In the Python scripts, *IndoorCore::CellSpace* in the UML is mapped as another Python class named *CellSpace*. Its attributes *sType*, *topClosure*, *sideClosure*, *bottomClosure* are directly transformed as four attributes with the same names. The value of *sType* is one of four: "ISpace", "sISpace", "sOSpace", and "OSpace", which corresponding to I-space, sI-space, sO-space, and O-space respectively. The values of the closures are decimals between 0 and 1. The *IndoorCore::CellSpaceBoundary* and its two attributes are converted as a Python class named *CellSpaceBoundary* with two attributes *bType* and *category*. The value of *bType* is either "Physical" or "Virtual", and value of *category* is "Top", "Side", or "Bottom". The (association) link is reflected by adding one more attribute named *containBoundary* to *CellSpace* and one attribute named *fromSpace* to the *CellSpaceBoundary*. The *containBoundary* contains the IDs of all *CellSpaceBoundary* that make up this *CellSpace*, while *fromSpace* records the ID of *CellSpace* where this *CellSpaceBoundary* comes from.

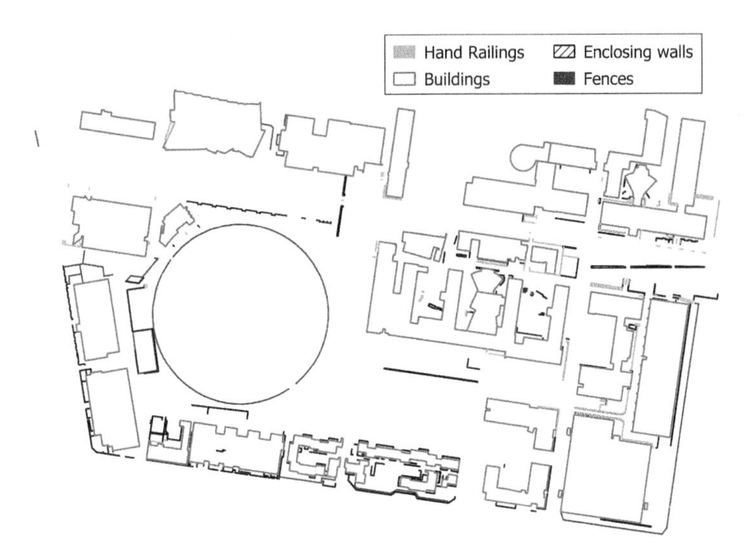

Figure 7.7 Physical boundaries used for space classification.

```
class CellSpaceBoundary:
  def __init__(self, bType, category, fromSpace):
  self.category = category
  self.bType = bType
  self.fromSpace = fromSpace
```

The following two tables shows the data structures of *CellSpace* and *CellSpace-Boundary* (Tables 4.2 and 4.3) in the computer memory. The names of the columns are the same as the attribute names of the two classes. Each *CellSpace* is represented as a quintuple: CellSpace = {sType, topClosure, sideClosure, bottomClosure, boundary}, while each *CellSpaceBoundary* as a triples: CellSpaceBoundary = {category, bType, space}. In the Table 4.2, “bi” (e.g., “b0”, “b3”)” are the IDs of *CellSpace-Boundary*, while “si” (e.g., “s1”, “s2”) in Table 4.3 are the IDs of *CellSpace*. Such IDs are not shown in the UML even the Python classes, but we consider each object will have an ID.

It should be noted that the above python classes and tables are designed for internal data management in this research. The UML model can be implemented as a spatial schema in a database management system (DBMS) covering all classes, data types and code lists. Using a DBMS will ensure management of large urban areas and entire cities.

We consider the roads and green areas as the two sources of sO-spaces and O-spaces. This section classifies them into semi-outdoor and outdoor based on their *sideClosure* (C^S), see Section 6.2.2. The footprints of buildings, hand railings, enclosing walls and fences are employed as physical boundaries (Figure 7.7).

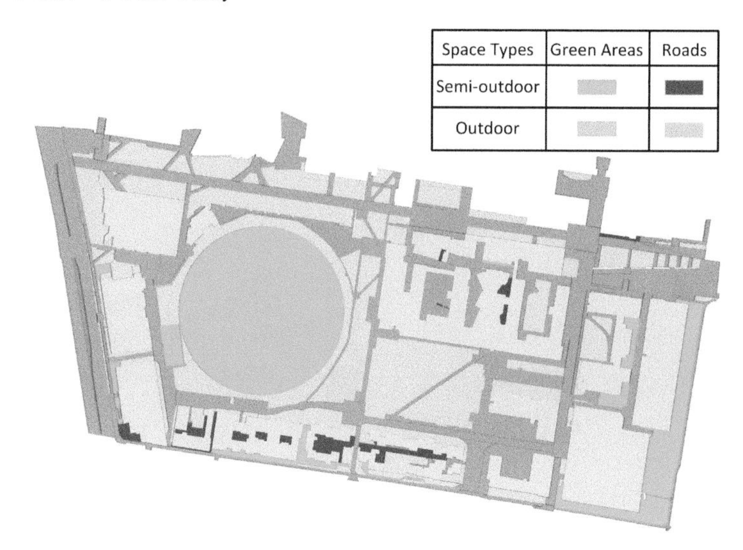

Figure 7.8 Classified road and green area footprints. The darker white polygons are building footprints.

The classified road and green spaces are shown in Figure 7.8. The results show that most roads and green areas are outdoor spaces, while yards and courtyards are sO-spaces. The classification of the footprints will be used as the semantic of the spaces reconstructed based on them, for instance, a semi-outdoor road footprint will be reconstructed into a sO-space in the following step.

Only the sI-spaces formed by the building with BIM model are reconstructed based on the approaches presented in Section 6.1. Other sI-spaces formed by shelters are reconstructed with an alternative way. The reconstructing process is extruding their projected footprints to 2 meters, which is similar to the reconstruction of building shells (Section 6.3). The reason why this alternative way is utilized is that the data source for most sI-spaces generation in this implementation is footprints on a 2D map, in which proper components of 3D built structure (such as *OuterCeilingSurface*, *OuterFloorSurface*) do not exist.

The sO-spaces and O-spaces are reconstructed on the basis of the procedures presented in Section 6.2. It should be noted that, in this implementation, the height of sO-spaces and O-spaces is set as 2 meters. However, height is variable. The reason why 2 meters is used is that we consider this height sufficient for pedestrian navigation. Furthermore, the purpose of setting height is enclosing spaces as 3D volumes. The tops of the spaces do not offer vertical constraints, for instance, if the height of a person is 2.5 meters, the spaces are still qualified for navigation.

The reconstructed spaces (Figure 7.9) show that all environments, where pedestrian navigation can happen are filled by spaces, except the physical boundaries. The data structures and classes used in this step are exactly the same as those presented in Section 7.2, but the *bottomClosure* of all spaces are assigned as 1.

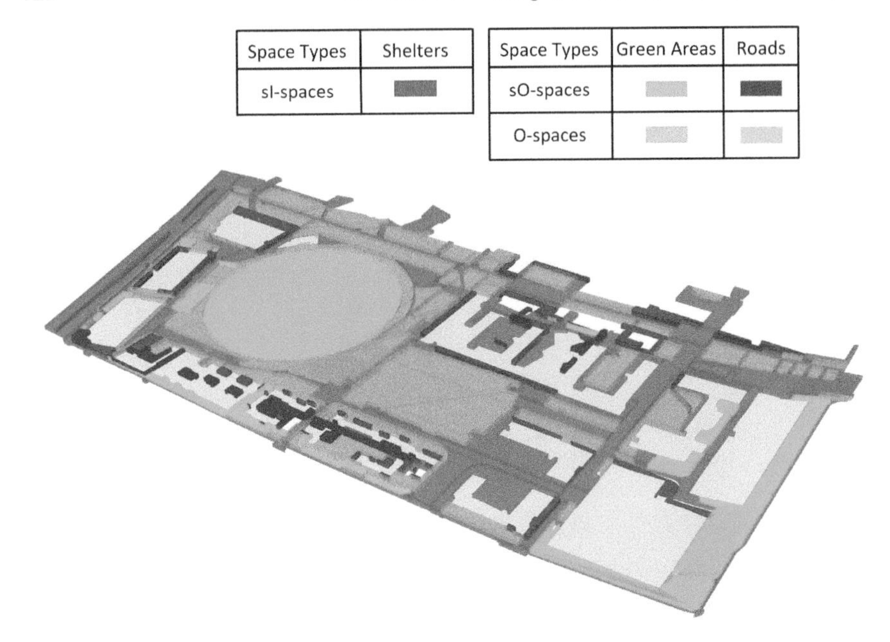

Figure 7.9 Reconstructed sI-spaces, sO-spaces, and O-spaces. The darker white polygons are building footprints.

7.3 SPACE SELECTION AND NAVIGATION NETWORK DERIVATION

Agents in this book are pedestrians who are 3D objects that have certain requirements for spaces. Thus, considering if spaces are large enough to accommodate pedestrians, the reconstructed spaces are selected based on their size. The minimum space needed by a pedestrian is estimated by Equation 1.3, i.e., $\{l',w',h'\} = \{750, 550, 1949.3\}$. Then, the space selection criteria for pedestrian navigation are: $r_S \geq 465.0mm$ and $h_S \geq 1949.3mm$. Only the qualified spaces are selected to participate in the navigation network derivation and navigation path planning. Nevertheless, it should be noted that the two criteria (r_S and h_S) only focus on the size of the space, which are the most basic criteria for space selection. Other than that, the space selection can consider the specific navigation purposes by adding other criteria, e.g., utilizing *topClosure* (C^T) as a criterion to select the spaces for the pedestrians who prefer to MTC-path or NSI-path during their navigation.

Indoor spaces are generally connected by doors (Figure 7.10(a)). Thus, theoretically, a navigation network can be derived from 3D spaces on the basis of Poincaré duality (Section 3.1.1). However, sI-spaces, sO-spaces, and O-spaces are naturally connected by virtual boundaries rather than doors (Figure 7.10(b)). Therefore, we use additional vertex on the face, where two spaces meet to indicate how to traverse the spaces. Even if the path does not intersect the spaces, the vertices are still needed, see the abstracted example (the green line in Figure 7.10(c)). Thus, curve edges and extra vertices are employed to force the navigation links to pass through the sharing

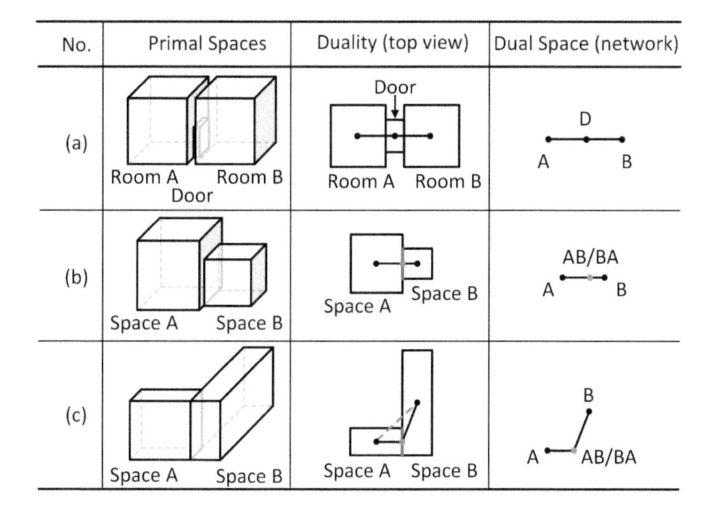

Figure 7.10 The duality used in this implementation. The red dots are extra vertices extracted from shared virtual boundaries.

of virtual boundaries. It should be mentioned that the computation of the additional vertices includes two steps: (i) evaluate if two spaces are sharing a virtual boundary surface based on 3D space intersection and (ii) extract the centroid of the intersection surface as the additional vertex. If there is more than one virtual boundary shared by two spaces, the vertex that ensures the shortest distance between the two spaces will be selected.

Furthermore, we integrated the I-spaces from geo-referenced BIM models based on the approach presented in [212] with all the reconstructed spaces. The integrated 3D spaces are shown in Figure 7.11. The data structure of the I-spaces is slightly

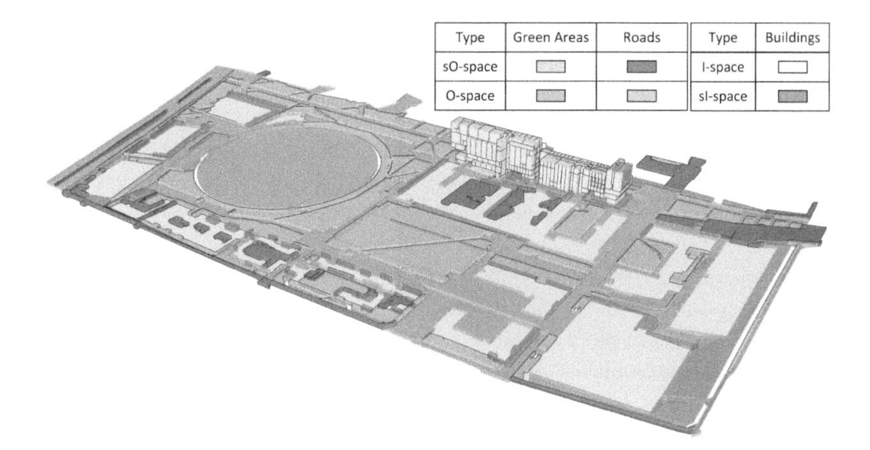

Figure 7.11 All the spaces in the selected area on the UNSW campus. The darker white polygons are building footprints.

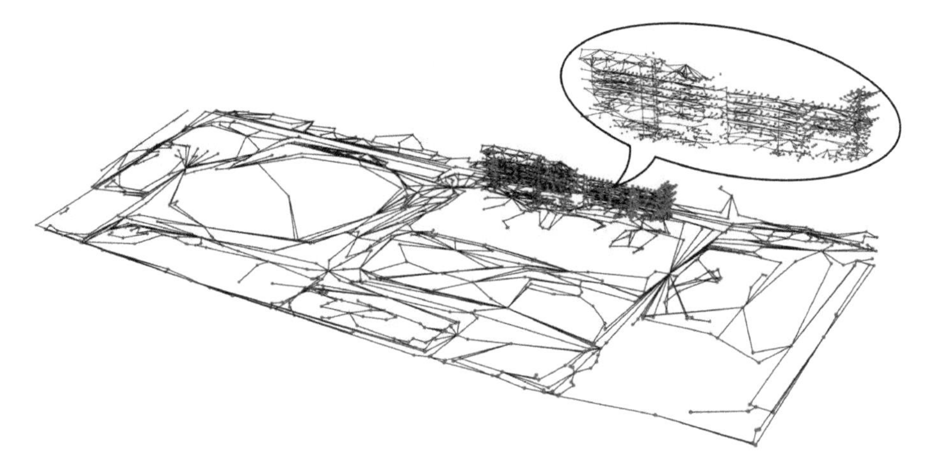

Figure 7.12 Derived unified navigation network in the test area. Black lines are edges and red dots are nodes.

different from other spaces, because we do not split the POLYHEDRALSURFACE of each I-space into top(s), side(s), and bottom(s). That is to say, the geometry of each space remains a solid. But, the *topClosure*, *sideClosure*, and *bottomClosure* of all I-spaces are assigned the value as 1. The name of each space corresponds to the "Ifc_name" column of the BIM model stored in an SQL database.

Utilizing all the selected I-spaces, sI-spaces, sO-space, and O-spaces, we derived the unified navigation network on the basis of Poincaré duality (Figure 7.12).

7.4 PATH PLANNING AND COMPARISON OF RESULTS

This section presents path planning, which includes two examples with five navigation paths. For comparison, OSM and Google Maps are also used for path planning.

7.4.1 EXAMPLES OF SEAMLESS NAVIGATION

Three navigation paths are performed to demonstrate the seamless navigation cases (Figures 7.13, 7.14, and 7.15). In the planned paths, black points are navigation nodes while the red are extra vertices. Blue curves indicate the navigation paths. Information about 3D spaces in the figure is linked to the table by number.

The first case is a path planning from an O-space (Road16) to a sO-space (Village green), see Figure 7.13, which is used to demonstrate the sO-spaces and the spaces without clear paths can be included in navigation path planning. The paths planned by OSM and Google Maps try to guide pedestrians to walk around the spaces without clear paths (such as green areas), which we call as detour issues since obviously pedestrians can walk through green areas directly to the destination. The path in our approach shows the detour issues are avoided by involving sO-spaces and O-spaces (greenArea25 and greenArea29) in the navigation map. We noticed that the detour

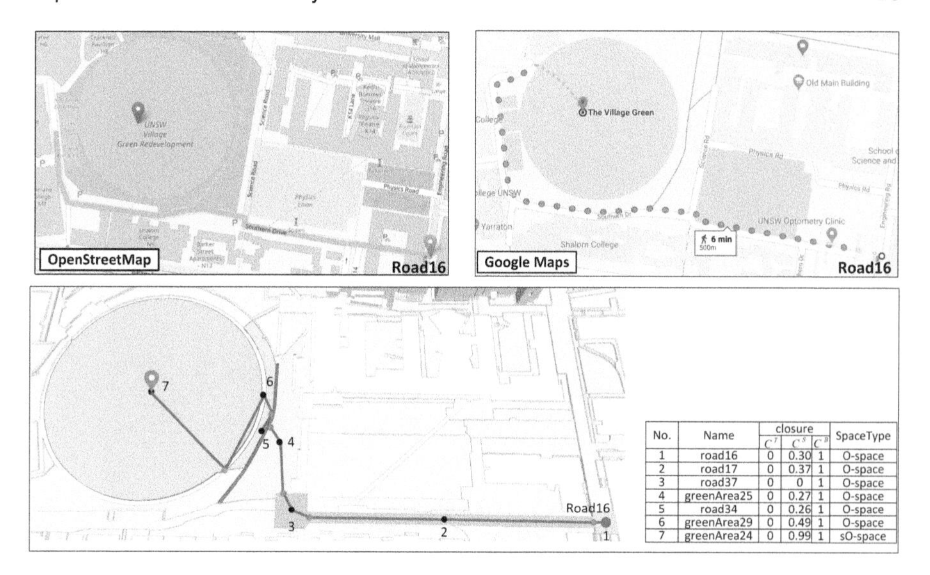

No.	Name	closure			SpaceType
		C^T	C^S	C^B	
1	road16	0	0.30	1	O-space
2	road17	0	0.37	1	O-space
3	road37	0	0	1	O-space
4	greenArea25	0	0.27	1	O-space
5	road34	0	0.26	1	O-space
6	greenArea29	0	0.49	1	O-space
7	greenArea24	0	0.99	1	sO-space

Figure 7.13 From O-space to sO-space (Road16 ⇝ Village green).

issues are avoided, but some new detour issues occur. For instance, it has undesirable zigzags between greenArea25 (node 4) to destination (node 7). Another issue is that partial segments of the navigation path are exposed to the outside of spaces. The reason why the two issues occur is the road space (road34 - node 5) and green area space (greenArea25 - node 6) are not convex polygons. The undesirable zigzags result from non-convex spaces. The undesirable zigzags are caused by non-convex spaces. One way to ensure that all spaces are convex polygons is space division [3]. But, this topic is not included in this research, but it has been listed in the future work, see Section 8.4.2 Apply Space Subdivision.

The second path planning is taking an O-space (Road19) as the departure while an I-spaces (sI-space47) as destination (Figure 7.14), which is designed to show that the sI-space can be departure/destination/transition in a navigation path. In the OSM and Google Maps, we cannot choose a sI-space as the destination since no clues show that somewhere is a sI-space. Our approach can indicate sI-spaces and includes them as departure, destination, or transition in the navigation map. Furthermore, the path shows the possibility to consider the vertical constraint (height bottleneck) in navigation. For instance, the sI-space (sI60), which formed by a building bridge. The accurate height of this bridge has a decisive effect on the navigation, specifically, if the height of a user is larger than this height, this path will be unavailable.

The third case is a path from a sO-space (Road80) to an I-space (Room 4036), see Figure 7.15, which is designed to demonstrate the seamless navigation. Considering we have only one BIM building that has I-spaces, if both departure and destination are located within this building, the navigation becomes indoor navigation. Thus, to demonstrate the seamless navigation, we chose a place out of the building as the departure while a place as the destination inside of the building. In OSM and Google

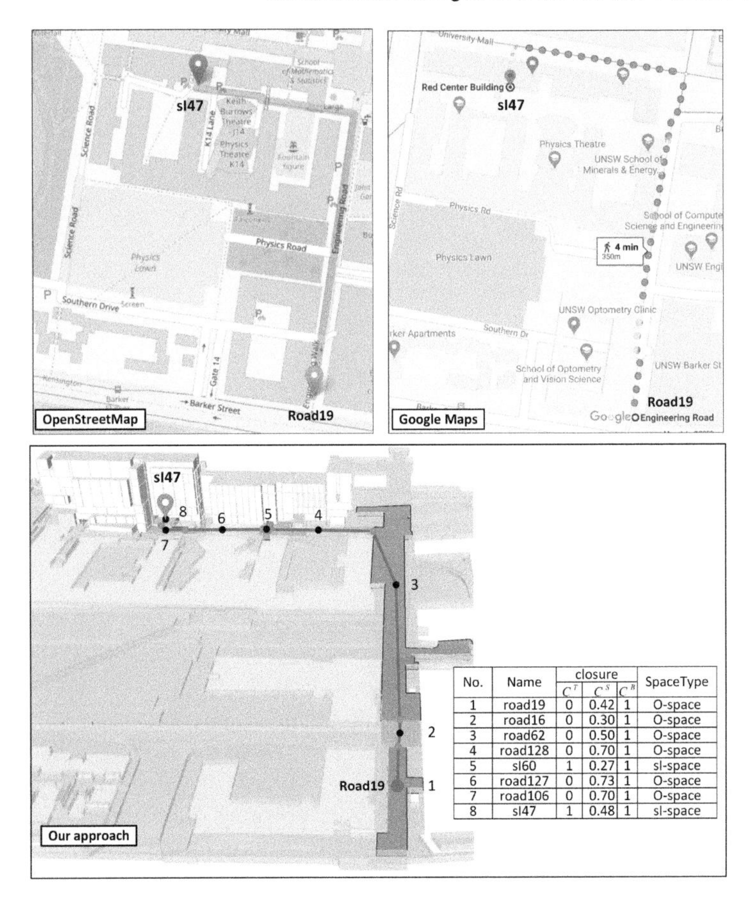

No.	Name	closure C^T	closure C^S	closure C^R	SpaceType
1	road19	0	0.42	1	O-space
2	road16	0	0.30	1	O-space
3	road62	0	0.50	1	O-space
4	road128	0	0.70	1	O-space
5	sl60	1	0.27	1	sl-space
6	road127	0	0.73	1	O-space
7	road106	0	0.70	1	O-space
8	sl47	1	0.48	1	sl-space

Figure 7.14 From O-space to sI-space (Road19 ⇝ sI-space47).

Maps, the navigation is limited in outdoor and there are no details of I-spaces. Thus, an I-space (room) cannot be chosen as the destination, and pedestrians are failed to get navigation paths in indoor parts. Our approach overcomes such kinds of shortcomings. On one hand, indoor spaces are able to be set as departures, destinations or transitions, and on the other hand, the navigation path includes the indoor parts. The path shows that seamless navigation is possible and there is no difference between indoor and outdoor from a navigation point of view.

7.4.2 EXAMPLE OF MTC-PATH & NSI-PATH

As mentioned in Chapter 5, the MTC-path (Section 5.2.2) and NSI-path (Section 5.2.3) are two new path options developed based on sI-space. Both their first step is the sI-space selection. Then, a new navigation network can be derived based on selected spaces. In this paths planning test, the navigation network derived in the last section is used as we assume that all the sI-spaces have a qualified *topClosure*

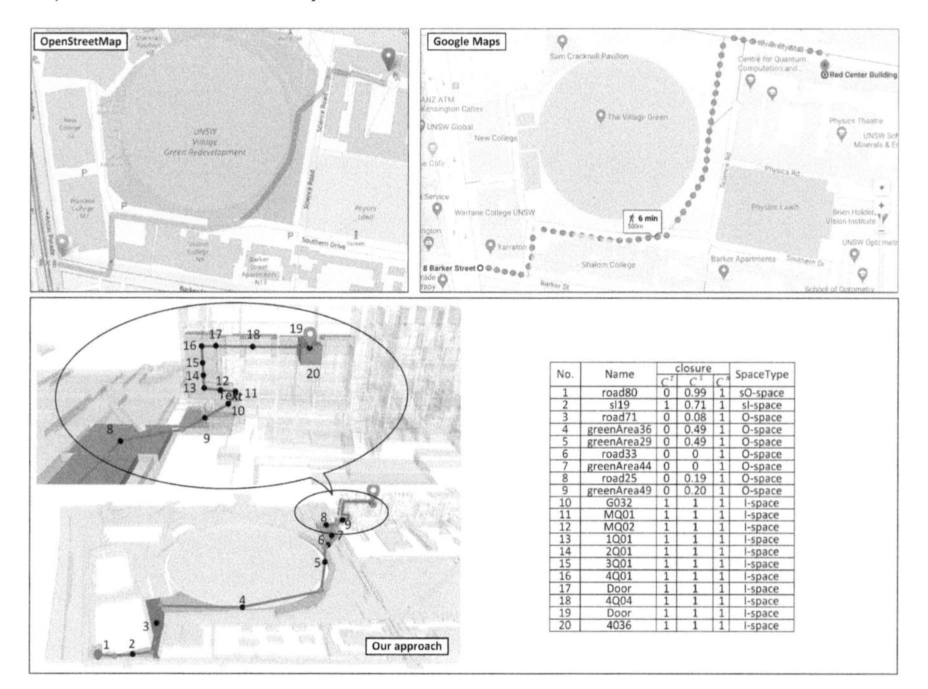

No.	Name	closure			SpaceType
		C^T	C^S	C^B	
1	road80	0	0.99	1	sO-space
2	sI19	1	0.71	1	sI-space
3	road71	0	0.08	1	O-space
4	greenArea36	0	0.49	1	O-space
5	greenArea29	0	0.49	1	O-space
6	road33	0	0	1	O-space
7	greenArea44	0	0	1	O-space
8	road25	0	0.19	1	O-space
9	greenArea49	0	0.20	1	O-space
10	G032	1	1	1	I-space
11	MQ01	1	1	1	I-space
12	MQ02	1	1	1	I-space
13	1Q01	1	1	1	I-space
14	2Q01	1	1	1	I-space
15	3Q01	1	1	1	I-space
16	4Q01	1	1	1	I-space
17	Door	1	1	1	I-space
18	4Q04	1	1	1	I-space
19	Door	1	1	1	I-space
20	4036	1	1	1	I-space

Figure 7.15 From sO-space to I-space (Road80 ⇝ Room 4036).

($C^T \geq 0.8$). The reason why we have such an assumption is that most sI-spaces are reconstructed from footprints of shelters, which makes us cannot estimate exact values of *topClosure* for them. The MTC-path and NSI-path are implemented with two navigation cases: A ⇝ B (Figure 7.16), C ⇝ D (Figure 7.17).

The first navigation case takes A as departure and B as the destination. The shortest path and MTC-path can be seen in Figure 7.16. All different approaches can offer the shortest path, but only our approach can offer the MTC-path and NSI-path. The navigation path computed by Google Maps obviously shows this system tries to guide pedestrians to go along the roads. Furthermore, the third dimension (vertical constraint) of outdoor spaces in OSM and Google Maps is neglected, but which can be considered in our approach, because the navigation network of our approach is derived from 3D spaces.

Our approach not only can offer the shortest path and MTC-path but also more detailed information of the path, such as covered/uncovered distance as well as the top-coverage-ratio. As shown in the figure at the bottom right, the navigation path includes as many sI-spaces as possible can increase the top-coverage-ratio of the path, see the black circled part in the MTC-path of Figure 7.16.

The second navigation case is from C (departure) to D (destination) (Figure 7.17). The path planning results of this case vary greatly. Path offered by OSM shows the pedestrians can go through a building. Navigation in Google Maps still tried to guide pedestrians to follow the roads. We expect our approach to have a similar shortest

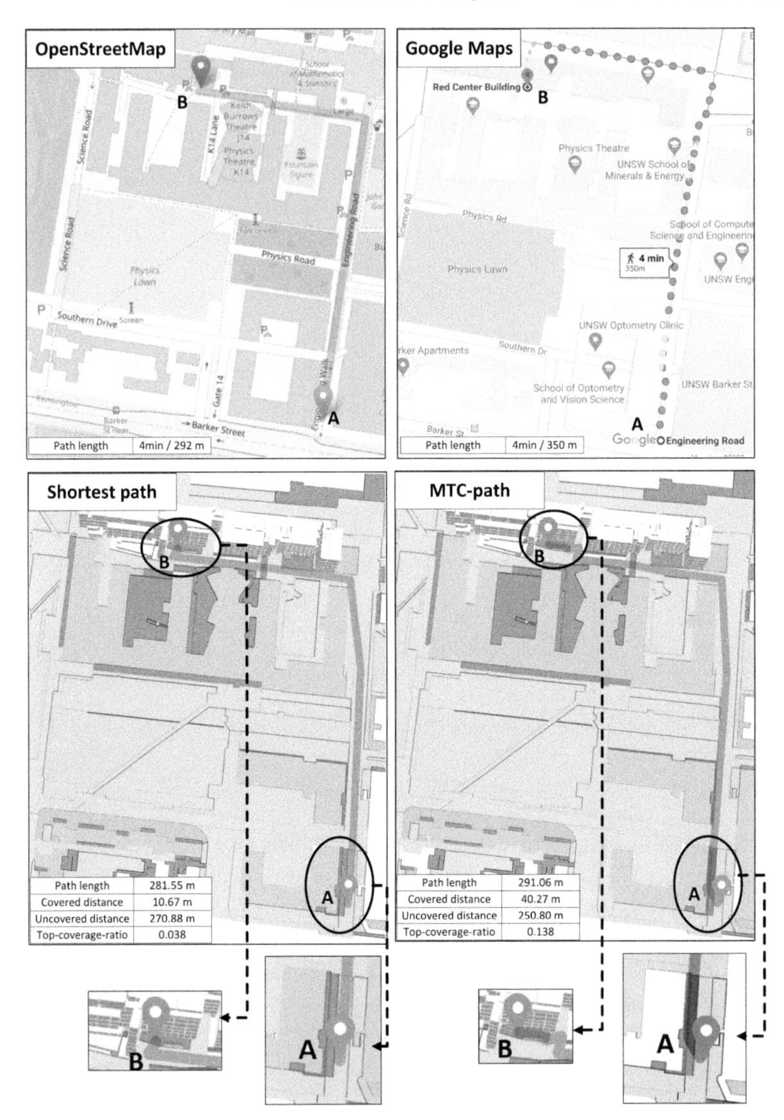

Figure 7.16 Navigation paths of A ⇝ B. Differences between shortest path and MTC-path are marked by black ellipses, which shows where the sI-spaces are involved.

path to the result from OSM, but with the reason that this building does not have BIM model and therefore is excluded, the shortest path of our approach slightly changes. In the MTC-path, it clearly shows that the path tries to include as many sI-spaces as possible to increase the covered distance and top-coverage-ratio of the path.

Based on the path selection strategy, both MTC-path in the two cases is recommended, since their uncovered distances are shorter than that of the corresponding shortest path (250.80 vs 270.88, and 270.48 vs 347.91) and the coverage ratio of the

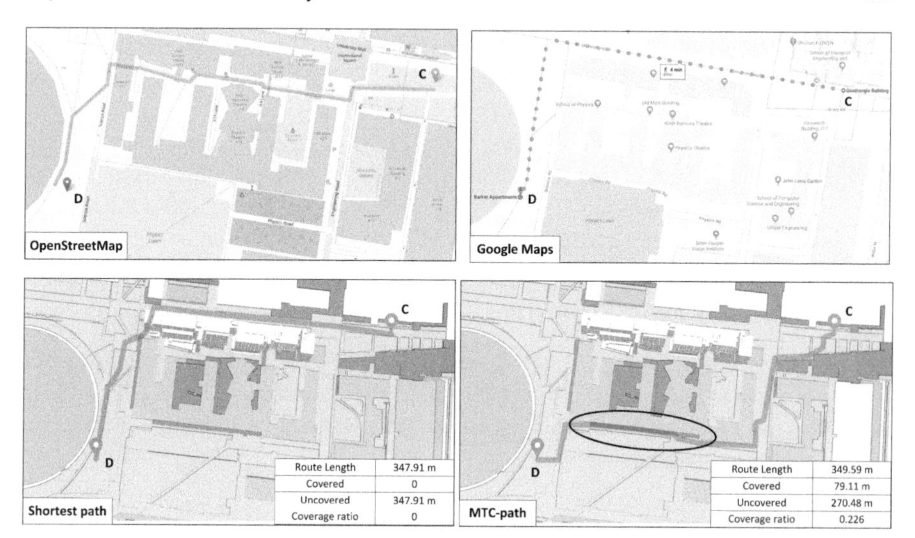

Figure 7.17 Navigation paths of C ⇝ D. The part marked by black ellipse shows where the sI-spaces are involved in MTC-path.

two MTC-paths is larger than that of the shortest paths (0.138 vs 0.038, and 0.226 vs 0). As mentioned above, the NSI-path is a compromise option when neither the shortest path nor MTC-path is recommended. Considering both MTC-path of two cases are recommended, this implementation does not further to compute the NSI-path. But, the NSI-path surely can be computed with our approach.

7.4.3 COMPARISON OF RESULTS

We compare the path planning results of OSM, Google Maps, and our approach from six aspects (Table 7.1): (i) if the sI-/sO-spaces are considered in the navigation map, (ii) it performs navigation in 2D or 3D, (iii) if it can compute the shortest path, (iv) if it can offer the seamless indoor/outdoor navigation path, (v) if it is able to offer the MTC-path, and (vi) if it can provide a NSI-path.

The results show that current navigation applications do not consider sI-spaces and sO-spaces in navigation map, perform navigation in 2D, and offer the shortest

Table 7.1
Comparisons of three navigation systems. × means no while ✓ means yes.

Approach	sI-/sO-space	2D/3D	Shortest path	Seamless path	MTC-path	NSI-path
OSM	×	2D	✓	×	×	×
Google Maps	×	2D	✓	×	×	×
Our approach	✓	3D	✓	✓	✓	✓

path as the only path option. However, thanks to the developed generic space definition framework, the unified 3D space-based navigation model, and reconstruction approach for 3D spaces, we can classify, defined, manage, structure, and reconstruct all types of spaces (especially sI-/sO-spaces) as 3D spaces and include them in the navigation map for seamless path planning. The navigation path shows that our approach performs navigation in 3D. More importantly, it not only can offer the shortest path but also the seamless navigation path, MTC-path, and NSI-path.

7.5 EXAMPLE OF ITSP-PATH

In order to enrich the experimental scene and simplify the data processing process, we used another indoor scene when testing and demonstrating the ITSP-path. In particular, we modeled one of the largest shopping centers in Sydney, Australia, for testing the path planning.

7.5.1 DATA PREPROCESSING

Two software packages were employed in this implementation: Quantum GIS (QGIS) and Rhinoceros (with Grasshopper). In the whole process, QGIS is used to edit the indoor maps, including editing footprints and attributes of indoor elements, and setting the coordinate system. Rhinoceros (with Grasshopper) is used to process the entire procedures of ITSP path planning, and the whole data process is developed in *Python* script (Figure 7.18). We edited the floor plan into four shapefile layers, including rooms, corridors, atriums, and POIs (Figure 7.19), in which except for the POIs, all other spaces are polygons. The shops, toilets, escalators, stairs, and lifts are included in the layer named Rooms, while the ATM and Bench are edited in POIs layer. Atriums and corridors are edited in two separated layers named Atriums and Corridor. Each element in each layer has three attributes: {ID, name, category}.

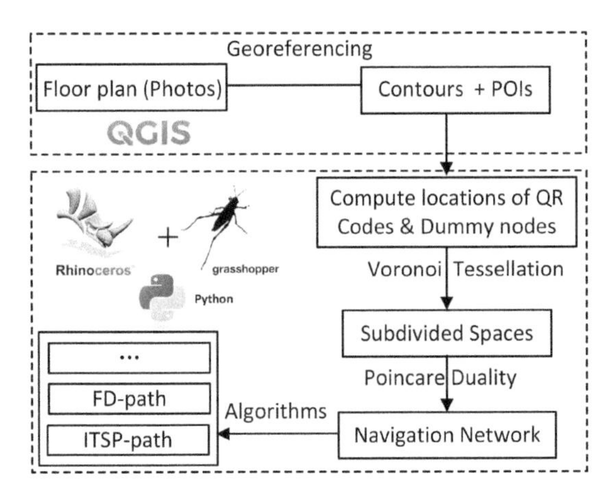

Figure 7.18 Workflow of the implementation and experiments.

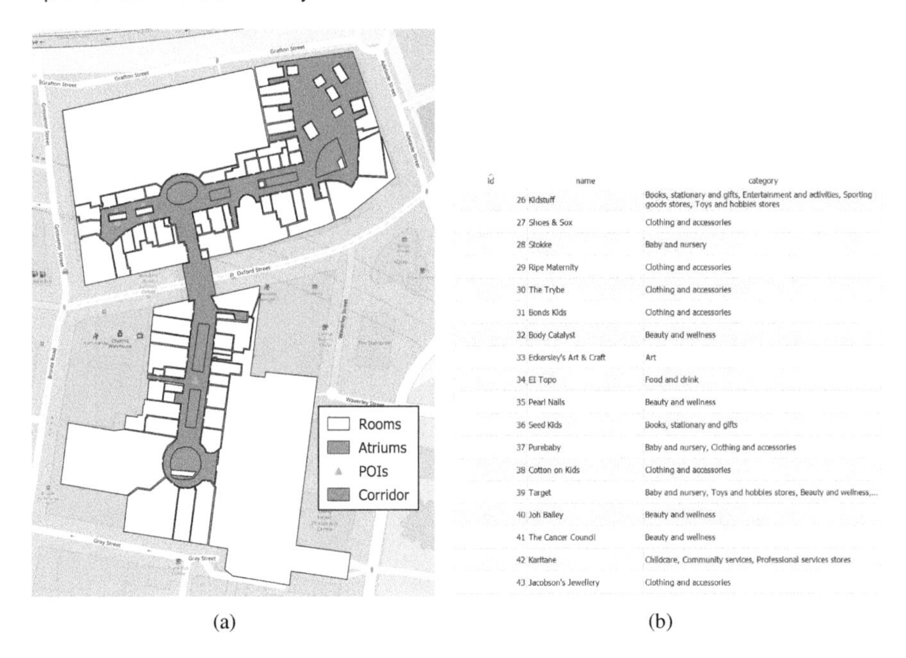

id	name	category
26	Kidstuff	Books, stationary and gifts, Entertainment and activities, Sporting goods stores, Toys and hobbies stores
27	Shoes & Sox	Clothing and accessories
28	Stokke	Baby and nursery
29	Ripe Maternity	Clothing and accessories
30	The Trybe	Clothing and accessories
31	Bonds Kids	Clothing and accessories
32	Body Catalyst	Beauty and wellness
33	Eckersley's Art & Craft	Art
34	El Topo	Food and drink
35	Pearl Nails	Beauty and wellness
36	Seed Kids	Books, stationary and gifts
37	Purebaby	Baby and nursery, Clothing and accessories
38	Cotton on Kids	Clothing and accessories
39	Target	Baby and nursery, Toys and hobbies stores, Beauty and wellness,...
40	Joh Bailey	Beauty and wellness
41	The Cancer Council	Beauty and wellness
42	Karitane	Childcare, Community services, Professional services stores
43	Jacobson's Jewellery	Clothing and accessories

(b)

Figure 7.19 Floor plan of the floor utilized for testing in the shopping mall. (a) The floor plan edited in QGIS; (b) The attributes of indoor elements.

ID records the id of each element, name keeps the semantic of the element (e.g., "Coles", "EscalatorL5_L4", "ToiletA_L5", "ATM_A_L5"), and category records the supplement information. For instance, the category of the shop named "Kidstuff", is "Books, stationary and gifts, Entertainment and activities, Sporting goods stores, Toys and hobbies stores".

The indoor spaces are rooms, corridor, POIs, obstacle, and atrium. Rooms are connecting to the corridor or other rooms by doors (which are simplified as lines), the POIs, atrium and obstacle are laying inside of the corridor, rooms/lifts or areas of stairs/escalators are also laying inside of the corridor but they are connecting with corridor only by the door (lines). By using QGIS, we edit the floor plan into four shapefile layers, including rooms, corridors, atriums, and POIs, in which except for the POIs, all other spaces are polygons. The shops, toilets, escalators, stairs, and lifts are included in the layer named Rooms, while the ATM and Bench are edited in POIs layer. Atriums and corridors are edited in two separated layers named Atriums and Corridor.

7.5.2 NAVIGATION NETWORK DERIVATION

Using indoor layers, locations of dummy nodes are computed (Figure 7.20(a)). Then, taking locations of QR code and dummy nodes as generators, the Voronoi tessellation is implemented to subdivide the corridors into non-overlapping polygons. After

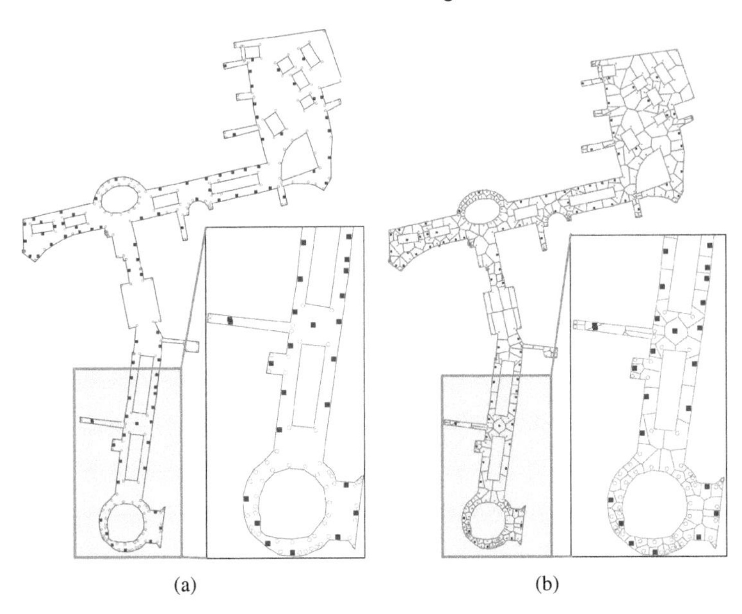

Figure 7.20 (a) QR codes and dummy nodes; (b) space subdivision based on locations of QR codes and dummy nodes. Black solid squares are locations of QR codes while the hollow circles are the locations of dummy nodes. Green lines are the split-lines

conducting region differences between the subdivided corridor polygons and the atriums/obstacles, we successfully obtain the final subdivided corridor (Figure 7.20(b)).

Hereafter, we implemented the Poincaré duality to automatically derive navigation networks. As mentioned in section 3.3.3, if a subdivided corridor space contains a QR code or dummy node, the location of the QR code or dummy node will be used as a node in navigation network to represent the space. Otherwise, the centroid of the spaces will be utilized as the node to represent the space. If two spaces are sharing a boundary, an edge is added to link the two nodes to indicate the two spaces are connected to each other. The things should be added are that the dummy nodes can nearly dispose all of the edges crossing corners of walls/obstacles/atriums, but there are still some exceptions. For instance, in Figure 7.21, one QR code and two dummy nodes are computed for the Escalator A space. On the basis of the three nodes, the space in front of the escalator is subdivided into space 1 (S_1), 2 (S_2), and 3 (S_3). Using Poincaré duality, there are three navigation edges will be derived to link Escalator A space, since each of the three spaces shares a boundary with escalator space (the parts marked by the red dotted line in Figure 7.21(a)). However, the edges between the escalator A space and S_1 and S_3 are incorrect as they are crossing the wall/atrium (the edges colored by red in Figure 7.21(b)). For this issue, we check and exclude such edges by identifying if they are intersecting with wall(s)/atrium(s), thereby having corrected networks (7.21(c)).

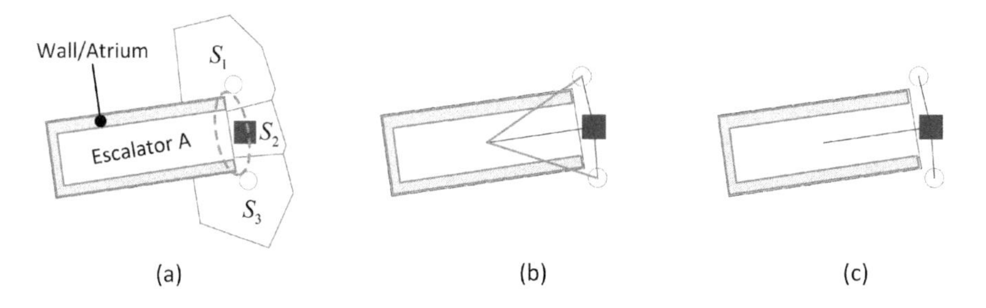

Figure 7.21 An example of exception case that dummy node cannot fully deal with. Black solid square is the location of QR code while the hollow circles are the locations of dummy nodes. (a) Subdivided spaces; (b) the derived navigation network; (c) the corrected navigation network.

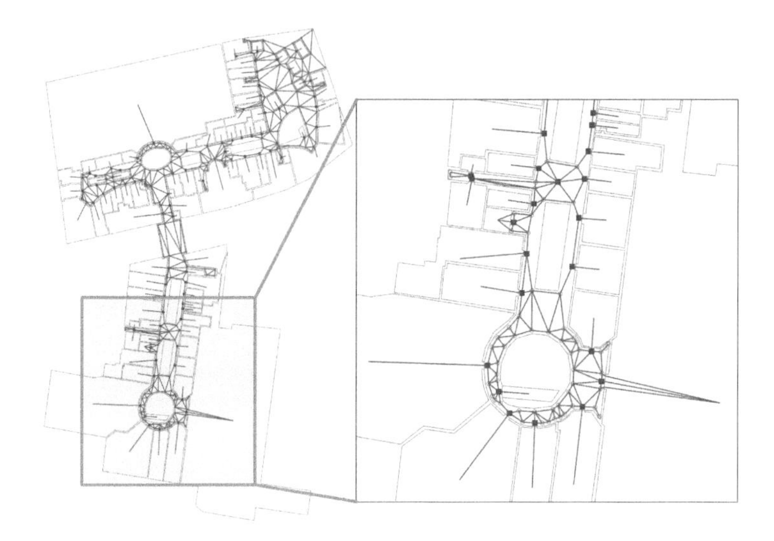

Figure 7.22 The derived navigation networks of Level 5. Black solid squares are locations of QR codes while the hollow circles are locations of dummy nodes. Blue lines are edges of navigation network.

The navigation network of level 5 (Figure 7.22) shows that all the QR codes are included as navigation nodes in navigation networks and dummy nodes successfully help edges avoid crossing corners of walls, obstacles, or atriums. All the indoor spaces and derived navigation networks of the shopping mall are shown in Figure 7.24. The distances between two floors are enlarged to 50 meters to make the spaces and navigation network more visible. In the navigation path planning, we estimate the direct distance between two adjacent levels is 3.5 meters. If the two levels are connected by stairs, the walking length is around 7 meters. Because, the acceptable slope of a stair is between 20° and 50°, and generally the preferred slope is between 30° and 35°. Suppose there are two connected stairs between two floors, the

Figure 7.23 The floor plan, QR codes, space subdivision and navigation network derivation of the selected floor. Black squares are QR codes, green lines in the left figure are split lines, blue lines in the right figure are edges of navigation network.

estimated waling path becomes 7 meters when taking 30° as the slope of stairs. Similarly, the walking length of using an escalator is also 7 meters when regarding its slope is 30°. The distance of the path length remains 3.5 meters when using lift.

During the navigation, pedestrians can scan the nearest QR codes to have their current locations. In this case study, QR codes are placed on the ground in front of the doors (the left figure of Figure 7.23). Then, locations of QR codes and the vertices of the corridor and atrium polygons are utilized as the generators of Voronoi Tessellation with envelope [182] to subdivide the corridor spaces. This approach can make sure each subspace only contains one QR code, and consequently, the subdivided spaces can indicate the topological relationship between QR codes, see the middle figure of Figure 7.23. Finally, on the basis of Poincaré duality, the topological relationships are automatically derived as indoor navigation networks (the right figure of Figures 7.23 and 7.24).

Figure 7.24 depicts all the floor plans and corresponding navigation network of the shopping mall for demonstration. In the figures, the distances between two floors are also enlarged to 100 meters to make the navigation network more visible. In computations of path length, this distance is adjusted based on the structures in the navigation path planning. In particular, we estimate the direct distance between two adjacent levels is 4.5 meters. If the two levels are connected by stairs, the walking length is around 9 meters. Because the acceptable slope of a stair is between 20 and 50 degrees, and generally the preferred slope is between 30 and 35 degrees. Suppose there are two connected stairs between the two levels, the estimated waling path becomes 9 meters when taking 30 degree as the slope of stairs (Figure 7.25(a)). Similarly, the walking length of using an escalator is also 9 meters when regarding its slope is 30 degree (Figure 7.25(b)). The distance of the path length remains 4.5 meters when using lift.

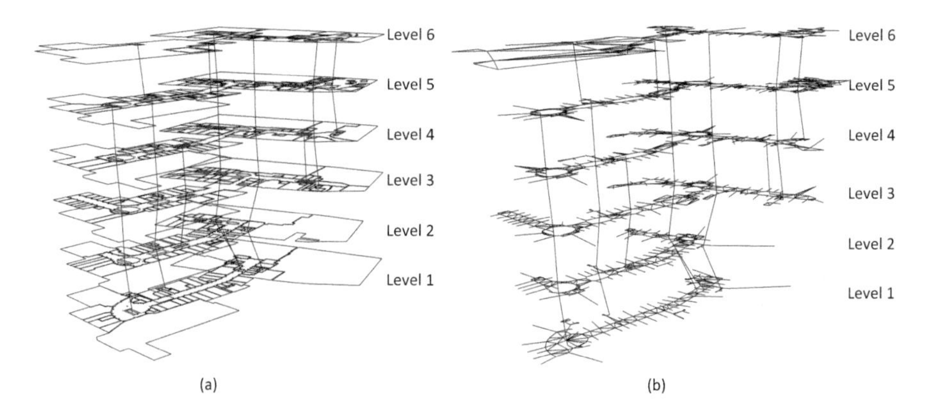

Figure 7.24 The shopping mall used for tests. (a) All the floor plans of the shopping mall, in which the links between different levels are lifts/escalators/stairs. (b) The derived navigation networks.

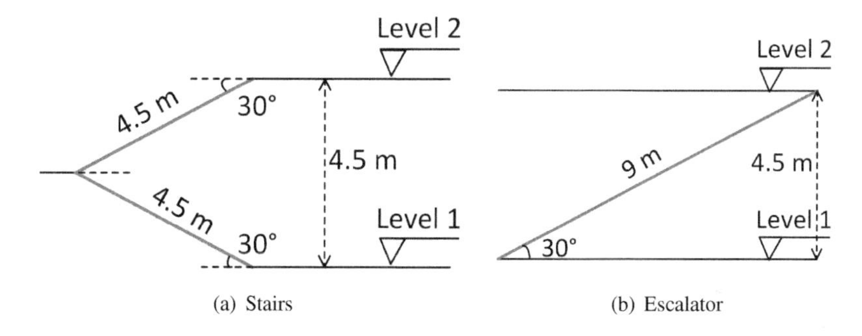

Figure 7.25 The distance estimation between two adjacent floors.

7.5.3 ITSP-PATH PLANNING

Two navigation cases in the shopping mall are utilized to demonstrate the ITSP-path planning. The first case is a pedestrian who plans to search some stuffs for kids on the fifth floor. From his/her current place ("LiftC_L5" or "EscalatorB_L5"), he/she would like to have a shortest navigation path that can help him/her to visit all the shops selling kids & babies stuff and then come back to the original place. This navigation can be translated as an ITSP-path planning case, in which the shops with stuff of babies or kids are desired destinations.

We firstly search the shops that have stuff of babies/kids as the desired destinations. Considering the name or category of shops can give the clues that they are selling goods for kids or babies, we conduct this process in the Query Builder of QGIS, in which we set a specific filter expression: "name" LIKE '%baby%' or "name" LIKE '%kid%' or "category" LIKE '%Baby%' or "category" LIKE '%kid%'. The search results are twelve shops (Table 7.2). Thus, the desired destinations are the

Table 7.2
The twelve shops with stuffs of babies or kids.

id	name	category
L5_6	Sheriden	Baby and nursery, Clothing and accessories, Home
L5_16	Adairs	Baby and nursery, Home
L5_21	Bed Bath N' Table	Clothing and accessories, Baby and nursery, Home
L5_22	Priceline	Baby and nursery, Discount and variety, Health and fitness
L5_25	Adairs Kids	Clothing and accessories, Home, Toys and hobbies stores
L5_26	Kidstuff	Books, stationary and gifts, Entertainment and activities, Sporting goods stores, Toys and hobbies stores
L5_28	Stokke	Baby and nursery
L5_31	Bonds Kids	Clothing and accessories
L5_36	Seed Kids	Books, stationary and gifts
L5_37	Purebaby	Baby and nursery, Clothing and accessories
L5_38	Cotton on Kids	Clothing and accessories
L5_39	Target	Baby and nursery, Toys and hobbies stores

twelve shops. Meanwhile, we suppose the costumer starts from LiftC_L5 and EscalatorB_L5 (Figure 7.26).

Having the twelve desired destinations and the navigation network, the two procedures that (i) compute navigation paths between every two destinations based on Dijkstra algorithm and (ii) set up a graph of destinations are conducted automatically. Then, taking the undirected graph as the input, the orders of departure and specified intermediate places are computed based on the B&B algorithm (Table 7.3).

The final step is combining the results as a navigation path (Figure 7.27). The length of the path starts from "LiftC_L5" is 1002.41 meters and that of the path from "EscalatorB_L5" is 1050.34 meters. These two navigation paths truly simulate the process of a customer coming out of the elevator or stairs to go shopping in stores that have products for kids or babies. Therefore, we believe ITSP-path is useful for such kinds of people.

The second navigation case is to show that the presented solution also can support ITSP-path planning in multi-floors. The scenario is a pedestrian who wants to visit several shops in the shopping mall (Figure 7.28). He/she starts from the supermarket named "Coles" on the first floor, which is the departure and final destination location of this indoor travel. The other desired destinations are: a shop named "Bondi Hair"

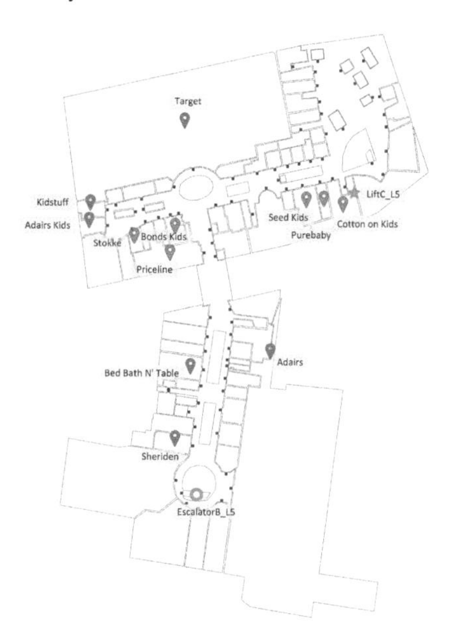

Figure 7.26 Departures and intermediate places (shops selling stuff of babies or kids).

that also on the first floor, "MacDonald's" on the third floor, and two shops ("Target" and "Myer L5") on the fifth floor.

In the path planning, we added a condition on the traveling ways of between different floors, i.e., lifts only, or all the ways (stairs, lifts, and escalators). The navigation paths computed by presented approach can be seen in Figure 7.29, in which the orders of the shops that will be visited in the two navigation paths are coincidentally

Table 7.3
TSP results with the two departure locations.

Departure	TSP results
LiftC_L5	LiftC_L5 ⇝ Seed Kids ⇝ Target ⇝ Kidstuff ⇝ Adairs Kids ⇝ Stokke ⇝ Bonds Kids ⇝ Priceline ⇝ Bed Bath N' Table ⇝ Sheriden ⇝ Adairs ⇝ Purebaby ⇝ Cotton on Kids ⇝ LiftC_L5
EscalatorB_L5	EscalatorB_L5 ⇝ Sheriden ⇝ Adairs ⇝ Seed Kids ⇝ Purebaby ⇝ Cotton on Kids ⇝ Target ⇝ Kidstuff ⇝ Adairs Kids ⇝ Stokke ⇝ Bonds Kids ⇝ Priceline ⇝ Bed Bath N' Table ⇝ EscalatorB_L5

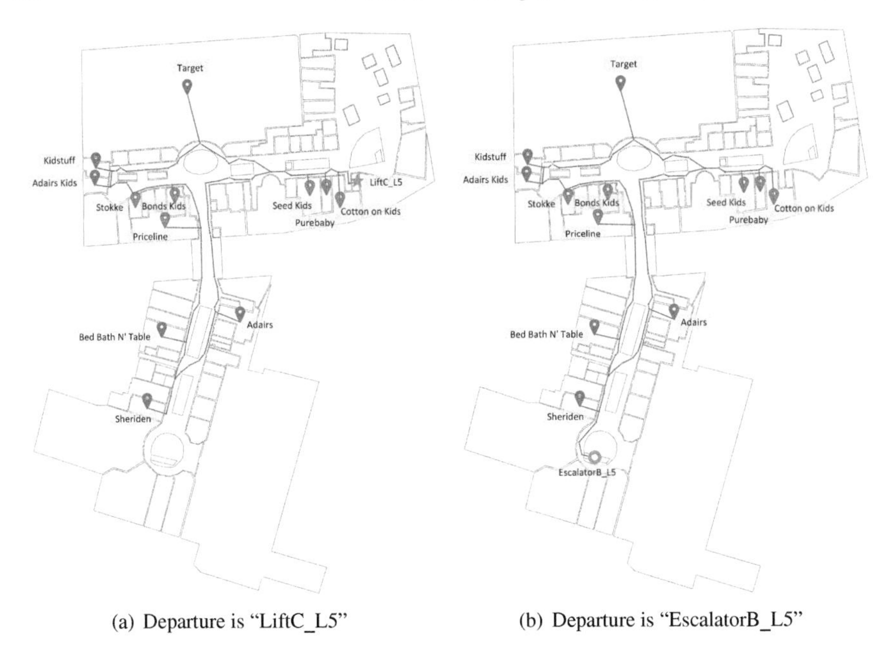

(a) Departure is "LiftC_L5" (b) Departure is "EscalatorB_L5"

Figure 7.27 The ITSP-path tests on the selected floor. The red star and circle represent the departure location, the blue location tags mark the shops that have stuff for kids/babies, the blue lines are the navigation paths, and the pink dots in the navigation path are the vertex of path segments.

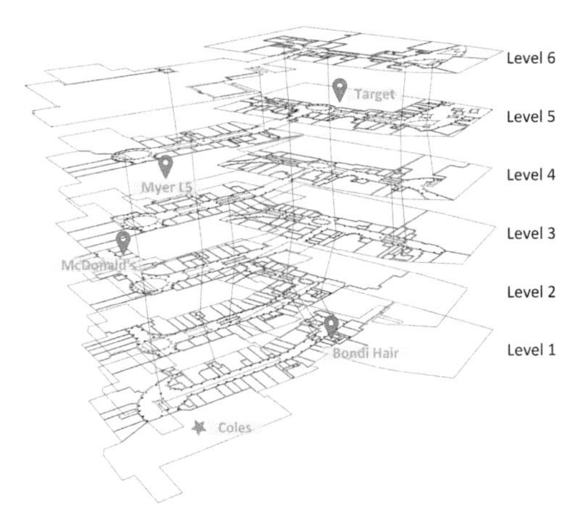

Figure 7.28 Departures and intermediate places in multi-floors.

the same: "Coles" ⇝ "Bondi Hair" ⇝ "Target" ⇝ "Myer L5" ⇝ "McDonald's" ⇝ "Coles". But, the two paths have different travel distance, 1202.06 meters for the using lifts (Figure 7.29(a)) only while 955.41 meters for vitalizing all types of ways (Figure 7.29(b)).

(a) Only Lifts (b) Stairs, escalators and lifts

Figure 7.29 The tests of ITSP-path on multi-floors. The red star represent the departure location, the blue location tags mark the shops that the customer is interested.

The two navigation paths show that the multi-floor ITSP-path planning based on the presented approach is feasible. During the navigation, pedestrians can have desired destinations that are distributed on the same or different floors. Meanwhile, the ways (stairs, lifts, or escalators) that the people using for transferring from one floor to another can be considered in ITSP-path planning.

7.6 SUMMARY

This chapter illustrates how to utilize all developed theories, models and approaches to perform indoor-outdoor seamless navigation path planning, including reconstruction of 3D spaces, implementation of the unified 3D space-based navigation model, derivation of navigation networks and path planning. There are several aspects should be further elaborated:

The height of sI-space should be strictly defined by the height of physical structures (e.g., roof, shelter, bridge). In this implementation, a contact height (2 meters) was employed, because (i) the input data for sI-space reconstruction are 2D footprints, thus information of physical tops of sI-spaces are not available and (ii) 2 meters are sufficient for the pedestrian navigation. For the similar reason, this contact height is also used in reconstruction of sO-spaces and O-spaces. But, it should be noted that the purpose of setting height is only enclosing sO-spaces and O-spaces as 3D volumes. In other words, this height of sO-spaces and O-spaces will not be used as the basis to determine whether such a space can meet the requirements of vertical constraints in navigation.

This presented approaches are based on 3D spaces, but it gives the impression that except for the I-spaces, the sI-spaces, sO-spaces, and O-spaces in the implementation are 2.5D. The reasons why we still regard the model as 3D are: (i) I-spaces employed in the model are 3D; (ii) sI-spaces should be 3D spaces generated from 3D models (e.g., CityGML LoD3). Due to lack of appropriate 3D models, sI-spaces are created by extruding 2D footprints up to a certain height; and (iii) all kinds of 3D space constructions take terrain into consideration to keep the topological consistency, which makes the navigation nodes are not on the same plane.

Space identification and subdivision is a pre-step of the navigation network derivation. A refined space subdivision would yield a smoother path. If the spaces are not subdivided properly, the navigation path will be undesirable. However, in the data preparation, only a rough space subdivision is conducted manually, which brings in some issues. For instance, because the spaces are not convex, partial zigzag, or navigation paths are exposed to the outside of spaces (Figure 7.13).

The navigation networks for path planning look still very rough (one space is abstracted as a node) and give the wrong impression that the developed model only can be based on 3D spaces for navigation network construction. It shows that they have the same structures with existing networks (i.e., NRG). Thence, although the unified model is based on 3D spaces, it remains valid if the environments are represented as 2D spaces. The information on the third dimension will be lost, but the navigation network will still be derived under unified rules and will, therefore, can combine with existing navigation networks and further provide seamless navigation.

The new path options can help pedestrians to have better navigation experiences. In this implementation, distance is the only factor that is considered when determining the optimal path between original position and destination. Other than the distance, there are other factors that may be also important, such as the pedestrian number, the crowd congestion, and the traffic time on the road. These factors are not considered in this book. Moreover, the navigation paths are a little angular, which is the common results of path planned based on the navigation network. Therefore, the navigation path should be further refined to avoid refraction.

8 Conclusion and Recommendations

This final chapter concludes the whole book. Then, the recommendations on seamless navigation research, developments, and implementations are presented as further research topics.

8.1 CONCLUSION

Contemporary public buildings are becoming conglomerates of open, semi-open and closed spaces, with indoor and outdoor sections, posing a set of challenges for humans to navigate seamlessly through such environments. Navigation systems generally rely on a network (nodes and edges) as an abstraction of underlying space connectivity. However, (i) indoor and outdoor networks are currently built differently. While indoor systems rely on indoor space subdivision approaches, current outdoor systems utilize road-based network approaches. (ii) Linking such networks via particular nodes is a possible but restrictive way to make a unified navigation model. (iii) Semi-bounded spaces in the built environment are not strictly indoor or outdoor spaces and are thus often omitted from navigation networks, further limiting navigation options. (iv) Current navigation networks are derived based on 2D-connectivity and the path planning assumes agents are points. Thus, vertical constraints during the navigation are not considered.

To overcome these challenges, based existing knowledge, this book reports new theories, models, and approaches to support seamless navigation path planning in all kinds of environments, which include a novel generic space definition framework, a unified 3D space-based navigation model, approaches for automatically reconstructing 3D spaces, and two new path options (MTC-path and NSI-path). The introduced novel generic space definition framework can categories and define the entire space formed by built structures as indoor outdoor, semi-indoor, and semi-outdoor spaces. The developed 3D space-based model illustrates that a same navigation network derivation and management approach for all types of spaces are possible. This elaborated procedures and algorithms demonstrate that all introduced spaces are able to be automatically derived as 3D volumes. The MTC-path and NSI-path prove that using the introduced spaces to define new navigation path options is feasible.

The following advances towards a seamless path planning are realized: (i) semi-bounded spaces are systematically and parametrically categorized as semi-indoor and semi-outdoor, thereby allowing to include them in navigation maps; (ii) all types of environments are uniformly modeled and managed as 3D spaces, which ensures the uniformity of the Dual Space, i.e., the navigation networks, thereby overcoming the difficulties of exchange and maintain the integrated navigation network; (iii) path planning based on 3D spaces can take into account vertical obstructions or

DOI: 10.1201/9781003281146-8

constraints; and (iv) the unified approach for space modeling and the extended space semantics opens possibilities to define and implement new navigation path.

This book illustrates that the spaces can be enriched with attributes and linked to supplementary data to derive user-dedicated information. Such extended information can be useful in many applications: shopping, facility management, guidance at airports and hospitals, etc. We believe this research contributes to LBS applications, urban data management, and urban analytic. For example, environmental parameters such as temperatures, noise or pollution can be analyzed more thoroughly.

Further elaboration and testing of the current work still need to be conducted: (i) investigate path navigation approaches to distinguish between different spaces and provide a path with respect to user preferences, e.g., the least-top-exposed path [59]; (ii) map the conceptual model to a DBMS spatial schema, compute paths based on the DBMS-supported navigation network, and investigate the storage and performance requirements of DBMS for near real-time path computation in large urban environments; (iii) test the computational efficiency and investigate abstraction/aggregation mechanisms to deal with city-scale environments and other types of data sources (e.g., BIM or point cloud data); (iv) address space accessibility by Land Administration Data Model (LADM) [213–215], which will allow us to build up a lightweight and dedicated navigation networks.

8.2 CONCLUSION ON TOPICS

This book covers three major topics and provides answers to six questions. The first topic is "Environments", which is the place where the navigation is performed; The second is "Space Representation", which deals with the digital representation of 3D spaces that define the environments; The last is "Unified Navigation Model and Path Options", which focusses on the management of 3D spaces and the computation of navigation paths.

8.2.1 ENVIRONMENTS

Q1: Which environments (spaces) should be considered in seamless navigation activities?

This research has shown that two more spaces should be introduced to perform seamless navigation, next to the commonly used Indoor and Outdoor. The semi-enclosed spaces, that can be found on the transition between indoor (I-space) and outdoor (O-space) should be considered in seamless navigation. This book provided formal definitions of the four spaces: I-space, O-space, sI-space and sO-spaces and demonstrated that with the four strictly defined spaces, all types of living environments can be covered for supporting seamless path planning. The developed conceptual space model ensures the automatic derivation of robust seamless navigation networks for path planning. Furthermore, including the sI-spaces and sO-spaces into the navigation opens new directions for further tailoring of the navigation path with respect to environmental factors, such as the MTC-path (Section 5.2.2).

Q2: Is it possible to classify and define all the complex environments as 3D spaces for seamless navigation?

This book has provided a definition of space. That is, for the purpose of navigation, our living environments can be defined in a unified way by 3D spaces. Section 2.4 has presented a generic space definition framework, which uniquely classifies and formally defines any space into one of the four categories: I-space, O-space, sI-space, and sO-space. Using this framework, all types of spaces are able to be included in the navigation model with the same management structures, thereby any application can rely on it to use the same network extraction approaches across the built environments and for providing a seamless navigation solution.

8.2.2 SPACES REPRESENTATION

Q3: Can we introduce unified names and terminology to define all kinds of spaces?

We have introduced three terms, *Top*, *Side*, and *Bottom*, to extend the basic building elements, roof, wall, floor, respectively for an indoor space to other structures in non-indoor spaces, such as the shelter, fence, and ground. That is, the three terms generalized the elements/structures in all types of spaces with generic notions, for instance, both roof and shelter can act as upper boundary of a space, but they are different. The former is mostly for I-space, while the latter for sI-space. Similarly, both wall and fence can be the surrounding boundary, and floor and ground can be the lower boundary (Section 1.3.2). The three terms allow all complex boundary configurations to be considered, although the materials, shapes, and dimensions of the structures that form the spaces are diverse. These three terms help to define all spaces uniformly (Section 2.4.1). Their geometry is also uniform, i.e., they are modeled as 3D volumes composed of *Top*, *Side*, and *Bottom*. This approach provides the basis for defining and developing a unified 3D space-based navigation model, i.e., network (Chapter 4).

Q4: What kind of terminology should be introduced to quantitatively define all kinds of spaces?

In addition to the three terms (*Top*, *Side*, and *Bottom*), we introduced six more notions to quantitatively define all kinds of spaces (Section 1.3.2), including *topClosure* (C^T), *bottomClosure* (C^B) and *sideClosure* (C^S), *Gradient* (G), *Physical boundary*, *Virtual boundary*. Furthermore, three threshold values (α, β and η) are introduced for C^T and two (γ and δ) for C^S to control their definitions (Section 2.4.2). As shown in implementation, these terms are able to help with classifying and identifying spaces for navigation uniquely and group them parametrically. In addition, spaces can be selected based on their closure for developing new path options. For example, the sI-spaces are selected based on their C^T for the path planning of MTC-path (Section 5.2.2) and NSI-path (Section 5.2.3).

8.2.3 UNIFIED NAVIGATION MODEL AND PATH OPTIONS

Q5: Is it possible to develop a 3D unified spatial model for seamless navigation?

The core of this book is a spatial model that can integrate all the defined parameters in a structured way to be able to be queried and modified with respect to changed environment and user needs. We developed a spatial model that can integrate the 3D spaces and their characteristics by extending the the concept of *CellSpace* and *CellSpaceBoundary* in IndoorGML (Chapter 4). The proposed modifications are relatively small and can be easily adopted by the standard, because all the classes remain the same as that in IndoorGML, and only several attributes are added. In particular, the modification for *CellSpace* is adding four attributes (sType, topClosure, sideClosure, and bottomClosure), while that of *CellSpaceBoundary* is two new attributes: category and bType.

With this unified 3D space-based model, we are able to (i) maintain all types of environments (indoor, outdoor, semi-indoor, and semi-outdoor) as 3D spaces, (ii) manage all types of spaces in a uniform way, such as management methods and network construction approach, and (iii) develop new seamless navigation path options based on the semantics of spaces.

Q6: Is it possible to develop new path options by considering the semantics of spaces?

Having defined all spaces formally, this book demonstrated that new path options, which differ from the commonly shortest/fastest can be defined. We put forward two roofed/sheltered navigation path options (MTC-path (Section 5.2.2) and NSI-path (Section 5.2.3)) based on the sI-spaces in the navigation map. This enriches the traditional navigation options (the shortest distance/time paths) and opens new directions for path options, which can bring in a new navigation experience for humans. With similar ideas, we can investigate new paths with I-spaces, even sO-spaces or O-spaces for other navigation preferences.

In simple terms, the new path options bring in three contributions to indoor-outdoor seamless navigation path planning: (i) people can have point-to-point navigation path with less undesirable detour issues; (ii) MTC-path and NSI-path are computed and suggested for users who need the shortest path with as many covers from the top as possible and (iii) the navigation performs in 3D, which aligns to the real situation that navigation is a process of pedestrians moving from one 3D space to another.

8.3 DISCUSSION

Aiming at incorporating all spaces into a navigation model to support seamless path planning, our research developed theories, models and algorithms. The test results illustrated that seamless navigation can successfully be achieved following space-based approach. Nevertheless, there are still some limitations that may affect the use of them in real navigation systems and need to be considered in further developments.

First of all, this research presented a generic definition framework, which is strictly based on parameters and thresholds. We attempted to assign values to them, such as the five thresholds ($\alpha = 0.05, \beta = 0.95$, $\gamma = 0.75$, $\delta = 0.95$, and $\eta = 0.6$) for the C^T and C^S in the generic definition framework. The threshold ($C^T \geq 0.8$) was utilized to select sI-spaces for the MTC-path and NSI-path. Theoretically, these parameters, thresholds and their suggested values are enough for the seamless path planning, because they can fully control classifications, definitions, and usages of the spaces. However, they are not verified by pedestrians in real scenarios. Thence, we cannot convincingly conclude that the values of them are well accepted by users. Therefore, to further verify this research, a set of tests should be conducted by deploying them in a navigation system or application. And then, improve them based on the feedback collected from pedestrians.

Secondly, a unified 3D space-based navigation model is developed for supporting space management. This model extends the two classes (*CellSpace* and *CellSpace-Boundary*) in IndoorGML, which ensures all types of spaces for navigation (indoor, semi-indoor, semi-outdoor, and outdoor) have the same representation, management methods, and network construction approach. But it is necessary to mention that there are still three limitations: (i) obstacles shall be represented and included in the model as non-navigable spaces. In particular, they shall be semantically categorized as static (e.g., pillar, tree), semi-mobile (e.g., small table, car), and mobile (e.g., fire) [37]. But, the model only focuses on the navigable spaces, thus the obstacles are overlooked; (ii) we presume all the spaces are available for navigation. However, some spaces might not be accessible. For instance, pedestrians are restricted from visiting some spaces, because they are private, occupied, not allowed, or closed. The accessibility may be a critical aspect in indoor and outdoor mixed contexts; (iii) the model is based on 3D spaces, but the 3D spaces might not be always available or needed to be created. It is possible to apply the same approach for 2D, but the information in the third dimension will be lost.

Thirdly, we mainly focus on the definition, classification, and reconstruction of the spaces formed by built structures. However, underground spaces, such as tunnels, landscape spaces, mines, caves, even desert, might be of interest for pedestrian navigation For instance, spaces can be formed by trees and some of these spaces may become non-navigable spaces or the treetops can form sI-spaces. Theoretically, such kind of spaces could also be classified and included in the unified navigation model and employed to enhance the MTC-path or NSI-path. However, such spaces a very different from built enclosures, which will require adaptation of parameters and thresholds. Therefore the framework, the approaches for 3D space reconstruction, and data sources should be further investigated.

Last but not the least, a pedestrian was utilized as the agent for the experiments. This allows us to narrow down the requirements according to which a space is qualified for navigation to (i) it is big enough to hold a pedestrian and (ii) its *topClosure* (C^T) is good enough for MTC-path and NSI-path planning purposes. However, pedestrians often can walk with extra equipment, such as trailers, child buggies, etc. Besides, other types of agents such as drones and robots can require seamless navigation. Such agents could have different requirements for space in size, C^T, as well as other aspects, such as the gradient of bottom. We expect the developed theories, models and approaches to be usable for theses different types of agents. Therefore, extensions of agent types could help to better test the whole research.

8.4 RECOMMENDATIONS FOR FURTHER RESEARCH

As discussed, several aspects still need to be investigated deeply to enhance the developed theories, models and approaches for indoor-outdoor seamless navigation path planning. Future work can be categorized into eight aspects, in which the first four directions are to strengthen and supplement the current results, while the remaining four are further sublimation of this research.

+ Extend the definition of spaces;
+ Apply space subdivision;
+ Include obstacles in the path planning;
+ Evaluate the navigation performance;
+ Popularize the results to other applications;
+ Reconstruct space based on other data sources;
+ Investigate space accessibility;
+ Develop and evaluate new navigation path options.

8.4.1 EXTEND THE DEFINITION OF SPACES

The generic definition framework developed by this research can systematically classify and define the built environment spaces into I-space, O-space, sI-space, and sO-space. Nevertheless, other than the spaces formed by built structures, pedestrians or other types of agents may also need navigation in the spaces formed by underground (e.g., caves), landscape spaces, mines, even desert. Therefore, it is essential to extend the generic space definition framework to cover such kind of spaces.

For extensions, the introduced terms by this book (Section 1.3.2), including *Top*, *Side*, *Bottom*, *topClosure*, *sideClosure*, *bottomClosure*, *Gradient*, *Physical/Virtual boundary*, still can be utilized, but the three threshold values (α, β and η) for C^T and the two (γ and δ) for C^S need to be re-estimated. For instance, landscape spaces are represented by boundaries with geometric representations, in which boundaries are called as surrounding planes, including ground plane, vertical plane, and overhead plane [183, 216]. As seen in Figure 2.11, treetops of arbors act as overhead planes, bushes as vertical planes, and ground/terrain as the ground plane. It means that the ground plane, vertical plane, and overhead plane can correspond to Bottom, Side, and Top, respectively. But the C^T and C^S of the spaces formed by trees cannot be computed based on the elaborated approaches, because it is very complex to estimate the physically enclosed areas and entire areas. Therefore, currently, it is hard to say how spaces formed by trees can be classified. For trees themselves, they are non-navigable while the spaces formed by them could be some of these four classified spaces, for instance, the spaces formed by treetops could be sI-spaces. That is, after additional investigation, the expectation is that forests can be included in *CellSpace* and *CellSpaceBounday*.

8.4.2 SPACE SUBDIVISION APPLICATION

Space subdivision is expected to bring in two benefits to path planning. One is to accurately subdivide the areas of outdoor to make sure spaces have the correct attributes, such as roads, streets, green areas, water bodies. Another is to help with deriving a (refined) navigation network, such as subdividing spaces into convex can make sure the navigation nodes stay within the spaces. In this book, we only have manually split roads/streets into closed polygons on the basis of their attributes, in which only buildings, shelters, roads, green areas, hand railings, enclosing walls, and fences are used as the borders. In the navigation network derivation, a road/street space is abstracted as a node without considering its convexity.

Thence, space subdivision rules or algorithms for automatically subdividing outdoor spaces based on different borders should be investigated. Particularly, drawing on the methods originally designed for indoor spaces can offer possibilities for extensions to outdoor, such as 3D indoor space subdivision based on the convex [37] or visibility [217]. Furthermore, consider more types of borders in outdoor spaces, because besides the borders mentioned above, outdoor spaces may have other natural borders, such as water, cliff, and traffic markers. If the agent becomes a vehicle, traffic markers especially should be considered as the borders, e.g., street with the road signs ("DO NOT ENTER") indicate that no cars are allowed to enter.

8.4.3 INCLUDE OBSTACLES IN PATH PLANNING

All parts in this research assume that the spaces are freely walkable and empty, i.e., there is no furniture/facilities. However, it is over idealistic, because this assumption fails to reflect the complexity of real indoor and outdoor environments thereby missing to consider the possibility that obstacles may change the navigation paths. Therefore, in future work, obstacles shall be included in the unified navigation model and further used for seamless navigation path planning. For instance, we can use the Flexible Space Subdivision framework (FSS) introduced by [37] to categories indoor obstacle spaces as static (e.g., pillar), semi-mobile (e.g., small table), and mobile (e.g., fire). As for the obstacles in sI-spaces, sO-spaces, and O-spaces, further investigations should be carried out on adapting the FSS framework or proposing new theories/approaches.

8.4.4 EVALUATE THE NAVIGATION PERFORMANCE

It is appropriate to say that the presented approach is suitable for precinct areas (e.g., a campus or similar), where pedestrians can reach on foot. However, navigation does not always only happen within such areas and the travel mode is not always limited to walk. For instance, someone may need a navigation path from an office at one end of the city to a friend's home at the other end. For such a case, a lot of spaces should be included and multiple locomotion modes may be needed. Due to the current implementations are performed based on files, once the scenario becomes a city level, without any abstraction/aggregation mechanisms, we may have great trouble to reconstruct the spaces, derive navigation networks, and plan the seamless navigation paths. More importantly, navigation paths are generally required to be planned in real-time and path planning becomes unacceptable if it takes too much time. Therefore, in the future work, two aspects should be conducted to cope with computing problems: (i) evaluate navigation path computational performance and involve multiple locomotion modes, and (ii) investigate abstraction/aggregation mechanisms for computing the spaces and deriving a network in large areas.

8.4.5 EXTEND THE RESULTS TO OTHER FIELDS

As mentioned in Section 1.3.1, agents can be broadly categorized as humans and robots, which have different locomotion modes and can vary in size and height, such as walking (pedestrians), rolling (trolleys, wheelchairs), flying objects (e.g., drones) or combinations of these. Therefore, other than pedestrian, research should be conducted on other types of agents, such as trolleys, wheelchairs, drones, or robots. Theoretically, the developed space classification, definition, the unified 3D space-based model, and space reconstruction approaches are still capable, but more parameters and criteria should be added to reflect the agents. For instance, compared to a pedestrian, a wheelchair user needs larger spaces in length and width but less space in the height, and he/she would require to be navigated with spaces that have no stairs and no undue bottom gradient.

Furthermore, this research developed the internal memory data structures (Section 7.2) by using the UML of the unified 3D space-based navigation model (Figure 4.1). This model is extended from IndoorGML and intended for representation indoor, semi-indoor, semi-outdoor and outdoor spatial information based on 3D spaces and network/graph. To make this model for multiple and multi-purpose uses, in the future, it is necessary to learn from IndoorGML, e.g., develop an XML-based schema of this model for the exchange of information. Other than that, another way to popularize this model is to utilize it as a data model for the persistent storage of data in the DBMS for different purposes.

8.4.6 RECONSTRUCT SPACES BASED ON OTHER DATA SOURCE

Space reconstruction is a key step of the presented unified model and path planning based on 3D spaces, because the 3D spaces might not be always available or needed to be created. No one currently known if existing data sources can provide these four navigation spaces at the same time. For instance, map providers or agencies generally cannot provide information of indoor and semi-indoor spaces. BIM models and footprints (2D maps) used in this research have been available from different sources, such as Estate Management, mapping agencies or OSM. If BIM models or 2D maps are not available, other algorithms could be the options, such as from point clouds [149]. Furthermore, with the development of 3D space modeling, we can access more 3D models, such as CityGML models. The on-going version of CityGML 3.0 is planning to model the outdoor as 3D spaces [172], which includes driving lane, sidewalks, even waterway [188].

Therefore, exploring 3D models as inputs to support the better 3D reconstruction of semi-outdoor and outdoor spaces can be a future direction. Other than 3D models, point clouds from airborne laser scans [149, 218, 219], or aerial photographs and terrestrial laser scans [220], or images from UAV [221] can also be the data sources for 3D reconstruction of spaces, although these kind of data currently are mainly for 3D (building) models generation [222].

8.4.7 INVESTIGATE SPACE ACCESSIBILITY

Currently, this research presumes that all the spaces are available for navigation. However, although the spaces are there, they might not be accessible. For instance, pedestrians are forbidden to visit some spaces because they are private, occupied, not allowed, or closed (Figure 8.1). Specifically, the space accessibility issues can be classified into property, object, legal, and time. The property issue means that spaces are private (e.g., the building with balconies in Figure 8.1(a)). There are two sI-spaces: one is formed by the roof and outer floor (balcony space), and the other is formed by the ground and the outer ceiling (space under the balcony). Generally speaking, the former cannot be used in the navigation model/system, as this space is private and only the house owner can use it, while the latter can be public. The object issue is that some spaces may be occupied by some objects. For instance, the spaces are occupied by plants (Figure 8.1(a)) or a car (Figure 8.1(b)). The legal issue usually arises in some public spaces, such as areas where pedestrians are not allowed to across some zones considering the security. The time issue means the spaces are available for pedestrians during certain time slots.

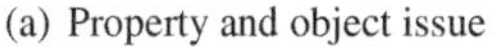

(a) Property and object issue

(b) Object issue

Figure 8.1 Example of space accessibility issues.

The Land Administration Domain Model (LADM) [213–215,223–226] can provide an option to solve the property issue, and the object issue can be addressed by considering the locations of objects. Based on the combined use of LADM models [213] with IndoorGML, indoor navigation can be performed with the accessibility of the indoor spaces based on user rights [227]. This practice is capable to make the navigation process more appropriate and simpler because the navigation path will avoid all of the non-accessible spaces based on the rights of the agents, which further can save computation resources for path planning in a large area. Therefore, in the future work, it is necessary to investigate how to add accessibility attribute to spaces by using the LADM.

8.4.8 DEVELOP AND EVALUATE NEW NAVIGATION PATH OPTIONS

This book developed two new navigation path options, MTC-path and NSI-path, in which the top-coverage-ratio of a path (P_{c_r}) is used as a metric for the optimal path determination. However, besides this, other metrics may also be interesting for some agent, such as the number of turns [228], safety [200], specific level of calories burn [201], minimum traffic related air pollution exposure [202], etc. Moreover, this research only attempted to develop new path options based on sI-space. It means that one of the future work is to develop new models that can consider more metrics, and another work can be developing new path options based on all types of spaces for further tailoring of the navigation path with respect to environmental factors. For instance, a path with as many I-spaces as possible, which could be very attractive in hot summer, since the I-spaces are generally equipped with air conditioners.

The experiences and feedback are critical for the research of seamless path planning, because the ultimate goal of developing new path options is to better navigate agents. However, currently, the new path options are merely devised by observations of human behaviors and needs in routing. Therefore, in future work, it is necessary to deploy the developed new path options in navigation systems, and then, evaluate and improve them based on the feedback of agents. For instance, by collecting questionnaires, investigating the acceptable C^T of space for selecting sI-spaces and I-spaces in MTC-path planning.

Papers Related to this Book

This book is organized based on publications of authors, which include seven published international journal papers, and three published international refereed conference papers.

1. **Yan, J.**, Lee, B.*, Zlatanova, S., Diakité, A.A., Kim, H. Navigation Network Derivation for QR Code-based Indoor Pedestrian Path Planning. *Transactions in GIS*, 2022, 26(3), 1240-1255. DOI: https://doi.org/10.1111/tgis.12912 (Journal)
2. **Yan, J.***, Zlatanova, S., Lee, B., Liu Q. Indoor Travelling Salesman Problem (ITSP) Path Planning. *ISPRS Int. J. Geo-Inf.* 2021, 10(9), 616. DOI: https://doi.org/10.3390/ijgi10090616 (Journal)
3. **Yan, J.***, Zlatanova, S., and Diakité, A.A. A unified 3D space-based navigation model for seamless navigation in indoor and outdoor. International Journal of Digital Earth. 2021, 14(8), 985-1003.DOI: https://doi.org/10.1080/17538947.2021.1913522 (Journal)
4. **Yan, J.***, Zlatanova, S., and Diakité, A.A. Two New Pedestrian Navigation Path Options based on Semi-indoor Space, *ISPRS Ann. Photogramm. Remote Sens. Spatial Inf. Sci.*, VI-4/W1-2020, 175–182. DOI: https://doi.org/10.5194/isprs-annals-VI-4-W1-2020-175-2020 (Conference)
5. Zlatanova, S., **Yan, J.***, Wang, Y., Diakité, A.A., Isikdag, U., Sithole, G., & Barton, J. Spaces in Spatial Science and Urban Applications – State of the Art Review. *ISPRS Int. J. Geo-Inf.* 2020, 9, 58. DOI: https://doi.org/10.3390/ijgi9010058 (Journal)
6. **Yan, J.***, Diakité, A.A., Zlatanova, S., & Aleksandrov, M. Finding Boundaries of Outdoor for 3D Space-based Navigation. *Transactions in GIS*. 2020, 24(2), 371–389. DOI: https://doi.org/10.1111/tgis.12613 (Journal)
7. **Yan, J.***, Diakité, A.A., Zlatanova, S. A generic space definition framework to support seamless indoor/outdoor navigation systems. *Transactions in GIS*. 2019, 23(6), 1273-1295. DOI: https://doi.org/10.1111/tgis.12574 (Journal)
8. **Yan, J.***, Diakité, A.A., Zlatanova, S., & Aleksandrov, M. Top-Bounded Spaces Formed by the Built Environment for Navigation Systems. *ISPRS Int. J. Geo-Inf.* 2019, 8, 224. DOI: https://doi.org/10.3390/ijgi8050224 (Journal)
9. **Yan, J.***, Zlatanova, S., Aleksandrov, M., Diakité, A.A., & Pettit, C. Integration of 3D objects and Terrain for 3D Modelling Supporting the Digital Twin. *ISPRS Annals of Photogrammetry, Remote Sensing & Spatial Information Sciences*, IV-4/W8, 2019, 147-154. DOI: https://doi.org/10.5194/isprs-annals-IV-4-W8-147-2019 (Conference)
10. **Yan, J.**, Diakité, A.A.*, & Zlatanova, S. An Extraction Approach of the Top-Bounded Space Formed by Buildings for Pedestrian Navigation.

ISPRS Annals of Photogrammetry, Remote Sensing & Spatial Information Sciences, IV-5, 2018, 247-254. DOI: https://doi.org/10.5194/isprs-annals-IV-4-247-2018 (Conference)

References

1. Andrew J May, Tracy Ross, Steven H Bayer, and Mikko J Tarkiainen. Pedestrian navigation aids: information requirements and design implications. Personal and Ubiquitous Computing, 7(6):331–338, 2003.

2. Marija Krūminaitė and Sisi Zlatanova. Indoor space subdivision for indoor navigation. In Proceedings of the Sixth ACM SIGSPATIAL International Workshop on Indoor Spatial Awareness, pages 25–31, 2014.

3. Sisi Zlatanova, Liu Liu, George Sithole, Junqiao Zhao, and Filippo Mortari. Space subdivision for indoor applications. In GISt Report No. 66. Delft University of Technology, OTB Research Institute for the Built Environment, 2014.

4. Michael Worboys. Modeling indoor space. In Proceedings of the 3rd ACM SIGSPATIAL International Workshop on Indoor Spatial Awareness, pages 1–6. ACM, 2011.

5. Hassan A Karimi, Ming Jiang, and Rui Zhu. Pedestrian navigation services: challenges and current trends. Geomatica, 67(4):259–271, 2013.

6. Alexandra Millonig and Katja Schechtner. Developing landmark-based pedestrian-navigation systems. IEEE Transactions on Intelligent Transportation Systems, 8(1):43–49, 2007.

7. Myat Thu Zar and Myint Myint Sein. Finding shortest path and transit nodes in public transportation system. In Genetic and Evolutionary Computing, pages 339–348. Springer, 2016.

8. Charlene Howard and Elizabeth K Burns. Cycling to work in phoenix: Route choice, travel behavior, and commuter characteristics. Transportation Research Record, 1773(1):39–46, 2001.

9. Claus Nagel, Thomas Becker, Robert Kaden, K Li, Jiyeong Lee, and Thomas H Kolbe. Requirements and space-event modeling for indoor navigation - how to simultaneously address route planning, multiple localization methods, navigation contexts, and different locomotion types. Open Geospatial Consortium, pages 1–54, 2010.

10. Ann Vanclooster, Nico Van de Weghe, and Philippe De Maeyer. Integrating indoor and outdoor spaces for pedestrian navigation guidance: A review. Transactions in GIS, 20(4):491–525, 2016.

11. Pekka Peltola and Terry Moore. Towards seamless navigation. In Multi-Technology Positioning, pages 125–147. Springer International Publishing, 2017.

12. Jinjin Yan, Abdoulaye A. Diakité, and Sisi Zlatanova. A generic space definition framework to support seamless indoor/outdoor navigation systems. Transactions in GIS, 23(6):1273–1295, 2019.

13. Taehoon Kim, Kyoung-Sook Kim, and Jiyeong Lee. How to extend indoorgml for seamless navigation between indoor and outdoor space. In International Symposium on Web and Wireless Geographical Information Systems, pages 46–62. Springer, 2019.

14. Liping Yang and Michael Worboys. A navigation ontology for outdoor-indoor space: (work-in-progress). In Proceedings of the 3rd ACM SIGSPATIAL international workshop on indoor spatial awareness, pages 31–34. ACM, 2011.

15. Ann Vanclooster and Philippe De Maeyer. Combining indoor and outdoor navigation: the current approach of route planners. In Advances in Location-Based Services, pages 283–303. Springer, 2012.

16. Joaquín Torres-Sospedra, Joan Avariento, David Rambla, Raúl Montoliu, Sven Casteleyn, Mauri Benedito-Bordonau, Michael Gould, and Joaquín Huerta. Enhancing integrated indoor/outdoor mobility in a smart campus. International Journal of Geographical Information Science, 29(11):1955–1968, 2015.

17. Jiantong Cheng, Ling Yang, Yong Li, and Weihua Zhang. Seamless outdoor/indoor navigation with wifi/gps aided low cost inertial navigation system. Physical Communication, 13:31–43, 2014.

18. Junjie Chen, Shuai Li, Donghai Liu, and Weisheng Lu. Indoor camera pose estimation via style-transfer 3d models. Computer-Aided Civil and Infrastructure Engineering, 37:335–353, 2022.

19. Hikmet Yucel, Rifat Edizkan, Taha Ozkir, and Ahmet Yazici. Development of indoor positioning system with ultrasonic and infrared signals. In 2012 International Symposium on Innovations in Intelligent Systems and Applications, pages 1–4. IEEE, 2012.

20. Ugur Yayan, Hikmet Yucel, et al. A low cost ultrasonic based positioning system for the indoor navigation of mobile robots. Journal of Intelligent & Robotic Systems, 78(3-4):541–552, 2015.

21. Masakatsu Kourogi, Nobuchika Sakata, Takashi Okuma, and Takeshi Kurata. Indoor/outdoor pedestrian navigation with an embedded gps/rfid/self-contained sensor system. Advances in Artificial Reality and Tele-Existence, pages 1310–1321, 2006.

22. Emidio Di Giampaolo. A passive-rfid based indoor navigation system for visually impaired people. In 2010 3rd International Symposium on Applied Sciences in Biomedical and Communication Technologies (ISABEL 2010), pages 1–5. IEEE, 2010.

23. Adam Satan. Bluetooth-based indoor navigation mobile system. In 2018 19th international carpathian control conference (ICCC), pages 332–337. IEEE, 2018.

24. David Ivan Vilaseca and Juan Ignacio Giribet. Indoor navigation using wifi signals. In 2013 Fourth Argentine Symposium and Conference on Embedded Systems (SASE/CASE), pages 1–6. IEEE, 2013.

25. Simon Robinson, Jennifer S Pearson, and Matt Jones. A billion signposts: Repurposing barcodes for indoor navigation. In Proceedings of the SIGCHI Conference on Human Factors in Computing Systems, pages 639–642, 2014.

26. Jussi Nikander, Juha Järvi, Muhammad Usman, and Kirsi Virrantaus. Indoor and outdoor mobile navigation by using a combination of floor plans and street maps. In Progress in Location-Based Services, pages 233–249. Springer, 2013.

27. Anahid Basiri, Pouria Amirian, and Adam Winstanley. The use of quick response (QR) codes in landmark-based pedestrian navigation. International Journal of Navigation and Observation, 2014:1–7, 2014.

28. S. N Shelke, Shinde Aboli, Unde Snehal, Yadav Damini, and Panbude Vishakha. Autonomous campus driver assistant for indoor and outdoor navigation. Imperial Journal of Interdisciplinary Research, 2(7):777–780, 2016.

29. Junho Park, Dasol Ahn, and Jiyeong Lee. Development of data fusion method based on topological relationships using indoorgml core module. Journal of Sensors, 2018:1–14, 2018.

30. Jiyeong Lee, Ki-Joune Li, Sisi Zlatanova, Thomas H. Kolbe, Claus Nagel, and Thomas Becker. OGC IndoorGML, Document No. 14-005r4. `http://docs.opengeospatial.org/is/14-005r5/14-005r5.html`, 2015. Accessed on January 26, 2022.

31. KyoHyouk Kim and John P Wilson. Planning and visualising 3d routes for indoor and outdoor spaces using cityengine. Journal of Spatial Science, 60(1):179–193, 2015.

32. Tee-Ann Teo and Kuan-Hsun Cho. Bim-oriented indoor network model for indoor and outdoor combined route planning. Advanced Engineering Informatics, 30(3):268–282, 2016.

33. Zhiyong Wang and Lei Niu. A data model for using openstreetmap to integrate indoor and outdoor route planning. Sensors, 18(7):2100, 2018.

34. Alexandra Millonig and K Schechtner. Decision loads and route qualities for pedestrians—key requirements for the design of pedestrian navigation services. In Pedestrian and Evacuation Dynamics 2005, pages 109–118. Springer, 2007.

35. Vadim Zverovich, Lamine Mahdjoubi, Pawel Boguslawski, Fodil Fadli, and Hichem Barki. Emergency response in complex buildings: Automated selection of safest and balanced routes. Computer-Aided Civil and Infrastructure Engineering, 31(8):617–632, 2016.

36. George Sithole and Sisi Zlatanova. Position, location, place and area: An indoor perspective. ISPRS Annals of Photogrammetry, Remote Sensing and Spatial Information Sciences, III-4:89–96, 2016.

37. Abdoulaye A. Diakité and Sisi Zlatanova. Spatial subdivision of complex indoor environments for 3d indoor navigation. International Journal of Geographical Information Science, 32(2):213–235, 2018.

38. Jinjin Yan, Abdoulaye A. Diakité, Sisi Zlatanova, and Mitko Aleksandrov. Finding outdoor boundaries for 3d space-based navigations. Transactions in GIS, 24(2):371–389, 2020.

39. Olena Klochkova, Wang Jia, Zena Wood, and Mike Worboys. Pedestrian route planning based on an enhanced representation of pedestrian network and probabilistic estimate of signal delays. In GIS Research UK (GISRUK), pages 1–6, 2017.

40. Simeon Andreev, Julian Dibbelt, Martin Nöllenburg, Thomas Pajor, and Dorothea Wagner. Towards realistic pedestrian route planning. In 15th Workshop on Algorithmic Approaches for Transportation Modelling, Optimization, and Systems (ATMOS 2015), pages 1–15. Schloss Dagstuhl-Leibniz-Zentrum fuer Informatik, 2015.

41. Jan Balata, Jakub Berka, and Zdenek Mikovec. Indoor-outdoor intermodal sidewalk-based navigation instructions for pedestrians with visual impairments. In International Conference on Computers Helping People with Special Needs, pages 292–301. Springer, 2018.

42. Jurgen Bodgan and Volker Coors. Using 3d urban models for pedestrian navigation support. In Proceedings of the ISPRS working group III/4, IV/8, IV/5: GeoWeb, pages 9–15. Citeseer, 2009.

43. Paulo Jorge Cambra, Alexandre Gonçalves, and Filipe Moura. The digital pedestrian network in complex urban contexts: A primer discussion on typological specifications. Finisterra, 54(110):155–170, 2019.

44. Guibo Sun, Chris Webster, and Xiaohu Zhang. Connecting the city: a three-dimensional pedestrian network of hong kong. Environment and Planning B: Urban Analytics and City Science, 48(1):60–75, 2021.

45. Anita Graser. Integrating open spaces into openstreetmap routing graphs for realistic crossing behaviour in pedestrian navigation. GI_Forum–Journal for Geographic Information Science, Salzburg, Österreich, July, pages 5–8, 2016.

46. Ernst Neufert, Vincent Jones, John Thackara, and Richard Miles. Architects' data. Granada St Albans, Herts, 1980.

47. Ernst Bosina, Mark Meeder, Beda Büchel, and Ulrich Weidmann. Avoiding walls: What distance do pedestrians keep from walls and obstacles? In Traffic and Granular Flow'15, pages 19–26. Springer, 2016.

48. Farlex, Inc. THE FREE DICTIONARY BY FARLEX. http://www.thefreedictionary.com/, 2003-2019. Accessed December 17, 2019.

49. Roman Trubka, Stephen Glackin, Oliver Lade, and Chris Pettit. A web-based 3d visualisation and assessment system for urban precinct scenario modelling. ISPRS Journal of Photogrammetry and Remote Sensing, 117:175–186, 2016.

50. Christof Beil and Thomas H Kolbe. Citygml and the streets of new york-a proposal for detailed street space modelling:(accepted). In Proceedings of the 12th International 3D GeoInfo Conference 2017, pages 9–16, 2017.

51. Hongchao Fan, Alexander Zipf, Qing Fu, and Pascal Neis. Quality assessment for building footprints data on openstreetmap. International Journal of Geographical Information Science, 28(4):700–719, 2014.

52. Keqi Zhang, Jianhua Yan, and S-C Chen. Automatic construction of building footprints from airborne lidar data. IEEE Transactions on Geoscience and Remote Sensing, 44(9):2523–2533, 2006.

53. Karim Hammoudi, Fadi Dornaika, and Nicolas Paparoditis. Extracting building footprints from 3d point clouds using terrestrial laser scanning at street level. ISPRS/CMRT09, 38:65–70, 2009.

54. Mathieu Brédif, Olivier Tournaire, Bruno Vallet, and Nicolas Champion. Extracting polygonal building footprints from digital surface models: A fully-automatic global optimization framework. ISPRS Journal of Photogrammetry and Remote Sensing, 77:57–65, 2013.

55. Permit Center. BUILDING HEIGHT CALCULATION INSTRUCTIONS. https://www.cob.org/documents/planning/applications-forms/building-height-calculation.pdf, 2002. Accessed December 17, 2019.

56. Collins. Collins Dictionary. https://www.collinsdictionary.com/, 2019. Accessed December 17, 2019.

57. Martin Brandli. Hierarchical models for the definition and extraction of terrain. Geographic Objects with Indeterminate Boundaries, 2:257, 1996.

58. Jay Lee. Comparison of existing methods for building triangular irregular network, models of terrain from grid digital elevation models. International Journal of Geographical Information System, 5(3):267–285, 1991.

59. Jinjin Yan, Abdoulaye A. Diakité, Sisi Zlatanova, and Mitko Aleksandrov. Top-bounded spaces formed by the built environment for navigation systems. ISPRS International Journal of Geo-Information, 8(5):224, 2019.

60. Liping Yang and Michael Worboys. Similarities and differences between outdoor and indoor space from the perspective of navigation. COSIT 2011, 2011.

61. Ki-Joune Li. Indoor space: A new notion of space. In International Symposium on Web and Wireless Geographical Information Systems, pages 1–3. Springer, 2008.

62. George A Miller. Wordnet: a lexical database for english. Communications of the ACM, 38(11):39–41, 1995.

63. Princeton University. Princeton WorldNetWeb. http://wordnetweb.princeton.edu/perl/webwn, 2010. Accessed December 17, 2019.

64. 2019 Lexico.com. OXFORD DICTIONARY. https://en.oxforddictionaries.com, 2019. Accessed December 17, 2019.

65. Merriam-Webster, Incorporated. Merriam-Webster DICTIONARY. https://www.merriam-webster.com/dictionary, 2019. Accessed December 17, 2019.

66. Dictionary.com, LLC. Dictionary.com. http://www.dictionary.com/, 2019. Accessed December 17, 2019.

67. Cambridge University Press 2019. Cambridge Dictionary. http://dictionary.cambridge.org, 2019. Accessed December 17, 2019.

68. Yoshinobu Ashihara. Exterior design in architecture. Van Nostrand Reinhold Company, 1981.

69. Jiyoeng Kim, Taeyeon Kim, and Seung-Bok Leigh. Double window system with ventilation slits to prevent window surface condensation in residential buildings. Energy and Buildings, 43(11):3120–3130, 2011.

70. Jiang He and Akira Hoyano. Measurement and evaluation of the summer microclimate in the semi-enclosed space under a membrane structure. Building and Environment, 45(1):230–242, 2010.

71. Kwangho Kim, Sanghyun Park, and Byungseon Sean Kim. Survey and numerical effect analyses of the market structure and arcade form on the indoor environment of enclosed-arcade markets during summer. Solar Energy, 82(10):940–955, 2008.

72. Nazliah Hani Mohd Nasir, Farha Salim, and Maheran Yaman. The potential of outdoor space utilization for learning interaction. pages 1–17. Kulliyyah of Architecture and Environmental Design, 4 2014.

73. Tzu-Ping Lin. Thermal perception, adaptation and attendance in a public square in hot and humid regions. Building and Environment, 44(10):2017–2026, 2009.

74. Jay Werb and Colin Lanzl. Designing a positioning system for finding things and people indoors. IEEE Spectrum, 35(9):71–78, 1998.

75. Christian S Jensen, Hua Lu, and Bin Yang. Graph model based indoor tracking. In 2009 Tenth International Conference on Mobile Data Management: Systems, Services and Middleware, pages 122–131. IEEE, 2009.

76. Christian S Jensen, Hua Lu, and Bin Yang. Indoor-a new data management frontier. IEEE Data Eng. Bull., 33(2):12–17, 2010.

77. Artur Baniukevic, Dovydas Sabonis, Christian S Jensen, and Hua Lu. Improving Wi-Fi based indoor positioning using bluetooth add-ons. In 2011 IEEE 12th International Conference on Mobile Data Management, volume 1, pages 246–255. IEEE, 2011.

78. Pengfei Zhou, Yuanqing Zheng, Zhenjiang Li, Mo Li, and Guobin Shen. Iodetector: A generic service for indoor outdoor detection. In Proceedings of the 10th ACM Conference on Embedded Network Sensor Systems, pages 113–126. ACM, 2012.

79. Artur Baniukevic, Christian S Jensen, and Hua Lu. Hybrid indoor positioning with wi-fi and bluetooth: Architecture and performance. In 2013 IEEE 14th International Conference on Mobile Data Management, volume 1, pages 207–216. IEEE, 2013.

80. Christian Kray, Holger Fritze, Thore Fechner, Angela Schwering, Rui Li, and Vanessa Joy Anacta. Transitional spaces: between indoor and outdoor spaces. In International Conference on Spatial Information Theory, pages 14–32. Springer, 2013.

81. Weiping Wang, Qiang Chang, Qun Li, Zesen Shi, and Wei Chen. Indoor-outdoor detection using a smart phone sensor. Sensors, 16(10):1563, 2016.

82. Jinjin Yan, Abdoulaye A. Diakité, and Sisi Zlatanova. An extraction approach of the top-bounded space formed by buildings for pedestrian navigation. ISPRS Annals of Photogrammetry, Remote Sensing & Spatial Information Sciences, 4(4), 2018.

83. Alessandro Mulloni, Daniel Wagner, Istvan Barakonyi, and Dieter Schmalstieg. Indoor positioning and navigation with camera phones. IEEE Pervasive Computing, 8(2):22–31, 2009.

84. Umit Isikdag, Sisi Zlatanova, and Jason Underwood. A BIM-oriented model for supporting indoor navigation requirements. Computers, Environment and Urban Systems, 41:112–123, 2013.

85. Sisi Zlatanova, Liu Liu, and George Sithole. A conceptual framework of space subdivision for indoor navigation. In Proceedings of the Fifth ACM SIGSPATIAL International Workshop on Indoor Spatial Awareness, pages 37–41. ACM, 2013.

86. Gavin Brown, Claus Nagel, Sisi Zlatanova, and Thomas H Kolbe. Modelling 3d topographic space against indoor navigation requirements. In Progress and New Trends in 3D Geoinformation Sciences, pages 1–22. Springer, 2013.

87. Liu Liu and Sisi Zlatanova. A two-level path-finding strategy for indoor navigation. In Intelligent Systems for Crisis Management, pages 31–42. Springer, 2013.

88. Nicholas A Giudice, Lisa Walton, and Michael F. Worboys. The informatics of indoor and outdoor space: a research agenda. In Proceedings of the 2nd ACM SIGSPATIAL International Workshop on Indoor Spatial Awareness, pages 47–53. ACM, 2010.

89. Ruey-Lung Hwang and Tzu-Ping Lin. Thermal comfort requirements for occupants of semi-outdoor and outdoor environments in hot-humid regions. Architectural Science Review, 50(4):357–364, 2007.

90. Van A. Timmeren and Michela Turrin. Case study 'the vela roof–unipol', bologna: use of on-site climate and energy resources. WIT Transactions on Ecology and the Environment, 121:333–342, 2009.

91. Twan Antonius Johannes van Hooff and Bert Blocken. Computational analysis of natural ventilation in a large semi-enclosed stadium. In Proceedings of the 5th European and African Conference on Wind Engineering, pages 1–11. Firenze University Press., 2009.

92. J Bouyer, J Vinet, P Delpech, and S Carré. Thermal comfort assessment in semi-outdoor environments: Application to comfort study in stadia. Journal of Wind Engineering and Industrial Aerodynamics, 95(9):963–976, 2007.

93. Xiao Hu Liu, Qiu Yu Chen, Hui Liu, Hui Yu, and Fei Yi Bie. Urban solar updraft tower integrated with hi-rise building – case study of wuhan new energy institute headquarter. In Solar Updraft Tower Power Technology, pages 67–71. Trans Tech Publications, 3 2013.

94. Leonardo M Monteiro and Marcia P Alucci. Transitional spaces in são paulo, brazil: mathematical modeling and empirical calibration for thermal comfort assessment. In Building Simulation 2007, pages 737–744, 2007.

95. Brian Clouston. Landscape Design with Plants. Newnes, 1977.

96. Norman K Booth. Basic elements of landscape architectural design. Waveland press, 1989.

97. Richard L. Austin. Elements of Planting Design. Number 04; SB472. 45, A8. 2002.

98. Almo Farina. Principles and Methods in Landscape Ecology: Towards a Science of the Landscape, volume 3. Springer Science & Business Media, 2008.

99. Stéphane Tonnelat. The sociology of urban public spaces. Territorial Evolution and Planning Solution: Experiences from China and France, pages 84–92, 2010.

100. Dafna Fisher-Gewirtzman. 3d models as a platform for urban analysis and studies on human perception of space. Usage, Usability, and Utility of 3D City Models–European COST Action TU0801, page 01001, 2012.

101. Tim R Oke. City size and the urban heat island. Atmospheric Environment (1967), 7(8):769–779, 1973.

102. Hongsuk H Kim. Urban heat island. International Journal of Remote Sensing, 13(12):2319–2336, 1992.

103. Ahmed Memon Rizwan, Leung YC Dennis, and LIU Chunho. A review on the generation, determination and mitigation of urban heat island. Journal of Environmental Sciences, 20(1):120–128, 2008.

104. Denise A Guerin. Issues facing interior design education in the twenty-first century. Journal of Interior Design, 17(2):9–16, 1991.

105. Buie Harwood. Comparing the standards in interior design and architecture to assess similarities and differences. Journal of Interior Design, 17(1):5–18, 1991.

106. Maureen Mitton and Courtney Nystuen. Residential Interior Design: A Guide to Planning Spaces. John Wiley & Sons, 2016.

107. Joo-Ho Lee and Hideki Hashimoto. Intelligent space—concept and contents. Advanced Robotics, 16(3):265–280, 2002.

108. Nobbir Ahmed and Harvey J Miller. Time–space transformations of geographic space for exploring, analyzing and visualizing transportation systems. Journal of Transport Geography, 15(1):2–17, 2007.

109. Liuqing Yang and Fei-Yue Wang. Driving into intelligent spaces with pervasive communications. IEEE Intelligent Systems, 22(1):12–15, 2007.

110. Fengzhong Qu, Fei-Yue Wang, and Liuqing Yang. Intelligent transportation spaces: vehicles, traffic, communications, and beyond. IEEE Communications Magazine, 48(11):136–142, 2010.

111. J-H Lee and Hideki Hashimoto. Intelligent space, its past and future. In IECON'99. Conference Proceedings. 25th Annual Conference of the IEEE Industrial Electronics Society (Cat. No. 99CH37029), volume 1, pages 126–131. IEEE, 1999.

112. Steve Wright and Alan Steventon. Intelligent spaces—the vision, the opportunities and the barriers. BT Technology Journal, 22(3):15–26, 2004.

113. Fei-Yue Wang, Daniel Zeng, and Liuqing Yang. Smart cars on smart roads: an ieee intelligent transportation systems society update. IEEE Pervasive Computing, 5(4):68–69, 2006.

114. Bin Liu, Fei-Yue Wang, Jason Geng, Qingming Yao, Hui Gao, and Buqing Zhang. Intelligent spaces: an overview. In 2007 IEEE International Conference on Vehicular Electronics and Safety, pages 1–6. IEEE, 2007.

115. Imad Afyouni, Cyril Ray, and Christophe Claramunt. A fine-grained context-dependent model for indoor spaces. In Proceedings of the 2nd ACM Sigspatial International Workshop on Indoor Spatial Awareness, pages 33–38. ACM, 2010.

116. Stephan Winter. Indoor spatial information. International Journal of 3-D Information Modeling (IJ3DIM), 1(1):25–42, 2012.

117. Urs-Jakob Rüetschi and Sabine Timpf. Modelling wayfinding in public transport: Network space and scene space. In International Conference on Spatial Cognition, pages 24–41. Springer, 2004.

118. Urs-Jakob Rüetschi. Wayfinding in scene space: Modelling Transfers in Public Transport. PhD thesis, University of Zurich, 2007.

119. Michela Turrin, Axel Kilian, Rudi Stouffs, and Sevil Sariyildiz. Digital design exploration of structural morphologies integrating adaptable modules: a design process based on parametric modeling. In Proceedings of Caad Futures 2009 International Conference. Joining languages, cultures and visions. Montreal, Canada, pages 17–19. CUMINCAD, 2009.

120. Chaya Chengappa, Rufus Edwards, Rajesh Bajpai, Kyra Naumoff Shields, and Kirk R Smith. Impact of improved cookstoves on indoor air quality in the bundelkhand region in india. Energy for Sustainable Development, 11(2):33–44, 2007.

121. Xiaoyu Du, Regina Bokel, and Andy van den Dobbelsteen. Building microclimate and summer thermal comfort in free-running buildings with diverse spaces: A chinese vernacular house case. Building and Environment, 82:215–227, 2014.

122. Tzu-Ping Lin, H Andrade, RL Hwang, S Oliveira, and A Matzarakis. The comparison of thermal sensation and acceptable range for outdoor occupants between mediterranean and subtropical climates. In Proceedings 18th International Congress on Biometeorology, Urban Climate, 2008.

123. Madhavi Indraganti. Adaptive use of natural ventilation for thermal comfort in indian apartments. Building and Environment, 45(6):1490–1507, 2010.

124. Giorgio Pagliarini and Sara Rainieri. Thermal environment characterisation of a glass-covered semi-outdoor space subjected to natural climate mitigation. Energy and Buildings, 43(7):1609–1617, 2011.

125. Taeyeon Kim, Shinsuke Kato, and Shuzo Murakami. Indoor cooling/heating load analysis based on coupled simulation of convection, radiation and hvac control. Building and Environment, 36(7):901–908, 2001.

126. Maria Philokyprou, Aimilios Michael, Stavroula Thravalou, and Ioannis Ioannou. Thermal performance assessment of vernacular residential semi-open spaces in mediterranean climate. Indoor and Built Environment, 27(8):1050–1068, 2017.

127. Shaogang Li. Users' behaviour of small urban spaces in winter and marginal seasons. Architecture and Behaviour, 10(1):95–109, 1994.

128. Aidan Slingsby and Jonathan Raper. Navigable space in 3d city models for pedestrians. In Advances in 3D Geoinformation Systems, pages 49–64. Springer, 2008.

129. Bin Cao, Maohui Luo, Min Li, and Yingxin Zhu. Thermal comfort in semi-outdoor spaces within an office building in shenzhen: A case study in a hot climate region of china. Indoor and Built Environment, 27(10):1431–1444, 2017.

130. Giorgio Pagliarini and Sara Rainieri. Dynamic thermal simulation of a glass-covered semi-outdoor space with roof evaporative cooling. Energy and Buildings, 43(2):592–598, 2011.

131. Jennifer Spagnolo and Richard De Dear. A field study of thermal comfort in outdoor and semi-outdoor environments in subtropical sydney australia. Building and Environment, 38(5):721–738, 2003.

132. Michela Turrin, Peter Von Buelow, Axel Kilian, and Rudi Stouffs. Performative skins for passive climatic comfort: A parametric design process. Automation in Construction, 22:36–50, 2012.

133. Danial Goshayeshi, Mohd Fairuz Shahidan, Farzaneh Khafi, and Ezzat Ehtesham. A review of researches about human thermal comfort in semi-outdoor spaces. European Online Journal of Natural and Social Sciences, 2(4):516, 2013.

134. Tzu-Ping Lin, Andreas Matzarakis, and Jia-Jian Huang. Thermal comfort and passive design strategy of bus shelters. In The 23rd Conference on Passive and Low Energy Architecture, Geneva, Switzerland, 6-8 September 2006. Citeseer, 2006.

135. Wei Yang, Nyuk Hien Wong, and Steve Kardinal Jusuf. Thermal comfort in outdoor urban spaces in singapore. Building and Environment, 59:426–435, 2013.

136. Alberto Ortiz, Francisco Bonnin-Pascual, Emilio Garcia-Fidalgo, and Joan P Company. Saliency-driven visual inspection of vessels by means of a multirotor. In The Workshop on Vision-Based Control & Navigation of Small, pages 20–46, 2015.

137. B. Amutha and Karthick Nanmaran. Development of a zigbee based virtual eye for visually impaired persons. In 2014 International Conference on Indoor Positioning and Indoor Navigation (IPIN), pages 564–574. IEEE, 2014.

138. Jorge Chen and Keith C Clarke. Indoor cartography. Cartography and Geographic Information Science, pages 1–15, 2019.

139. Srinivas Raghothama and Vadim Shapiro. Boundary representation deformation in parametric solid modeling. ACM Transactions on Graphics (TOG), 17(4):259–286, 1998.

140. Jaroslaw R Rossignac. Constraints in constructive solid geometry. In Proceedings of the 1986 Workshop on Interactive 3D Graphics, pages 93–110. ACM, 1987.

141. C.V. Ramakrishnan, S. Ramakrishnan, Ashish Kumar, and Marina Bhattacharya. An integrated approach for automated generation of two/three dimensional finite element grids using spatial occupancy enumeration and delaunay triangulation. International Journal for Numerical Methods in Engineering, 34(3):1035–1050, 1992.

142. B Gorte, M Aleksandrov, and S Zlatanova. Towards egress modelling in voxel building models. ISPRS Annals of Photogrammetry, Remote Sensing and Spatial Information Sciences, pages 43–47, 2019.

143. Mitko Aleksandrov, Sisi Zlatanova, Laurence Kimmel, Jack Barton, and Ben Gorte. Voxel-based visibility analysis for safety assessment of urban environments. ISPRS Annals of Photogrammetry, Remote Sensing and Spatial Information Sciences, pages 11–17, 2019.

144. Fangyu Li, Sisi Zlatanova, Martijn Koopman, Xueying Bai, and Abdoulaye Diakité. Universal path planning for an indoor drone. Automation in Construction, 95:275–283, 2018.

145. Ben Gorte, Sisi Zlatanova, and Fodil Fadli. Navigation in indoor voxel models. ISPRS Annals of Photogrammetry, Remote Sensing and Spatial Information Sciences, IV-2/W5:279–283, 2019.

146. Edward J Powers, Doug Gray, and Richard C Green. Artificial Vision: Image Description, Recognition, and Communication. Academic Press, 1996.

147. Thomas Becker, Claus Nagel, and Thomas H. Kolbe. A multilayered space-event model for navigation in indoor spaces. In Proceedings of the 3rd International Workshop on 3D Geo-Information, Seoul, Korea. Lecture Notes in Geoinformation & Cartography, pages 61–77. Springer, 2009.

148. Pawel Boguslawski and Christopher Gold. Construction operators for modelling 3d objects and dual navigation structures. 3D Geo-Information Sciences, pages 47–59, 2009.

149. O.B.P.M. Rodenberg, Edward Verbree, and Sisi Zlatanova. Indoor A* pathfinding through an octree representation of a point cloud. ISPRS Annals of the Photogrammetry, Remote Sensing and Spatial Information Sciences, 4:249, 2016.

150. James R Munkres. Elements of Algebraic Topology, volume 2. Addison-Wesley Menlo Park, 1984.

151. Filippo Mortari, Sisi Zlatanova, Liu Liu, and Eliseo Clementini. “improved geometric network model” (ignm): A novel approach for deriving connectivity graphs for indoor navigation. ISPRS Annals of the Photogrammetry, Remote Sensing and Spatial Information Sciences, 2(4):45, 2014.

152. Yueyong Pang, Chi Zhang, Liangchen Zhou, Bingxian Lin, and Guonian Lv. Extracting indoor space information in complex building environments. ISPRS International Journal of Geo-Information, 7(8):321, 2018.

153. José Nuno Beirão, André Chaszar, et al. Convex-and solid-void models for analysis and classification of public spaces. In Proceedings of the 19th CAADRIA International Conference, Kyoto, pages 253–262. CUMINCAD, 2014.

154. José Nuno Beirão, André Chaszar, and Ljiljana Cavić. Analysis and classification of public spaces using convex and solid-void models. In Future City Architecture for Optimal Living, pages 241–270. Springer, 2015.

155. Rusne Sileryte, Ljiljana Cavic, and José Nuno Beirão. Automated generation of versatile data model for analyzing urban architectural void. Computers, Environment and Urban Systems, 66:130–144, 2017.

156. Liu Liu. Indoor Semantic Modelling for Routing: The Two-level Routing Approach for Indoor Navigation. PhD thesis, 2017.

157. Marija Krūminaitė. Space subdivision for indoor navigation. Master's thesis, Dissertação (Mestrado em Geomática). Delft University of Technology, 2014.

158. Jiyeong Lee. A spatial access-oriented implementation of a 3-d gis topological data model for urban entities. Geoinformatica, 8(3):237–264, 2004.

159. Martijn Meijers, Sisi Zlatanova, and Norbert Pfeifer. 3d geoinformation indoors: structuring for evacuation. In Proceedings of Next Generation 3D City Models, volume 6, pages 11–16. Germany Bonn, 2005.

160. Bernhard Lorenz, Hans Jürgen Ohlbach, and Edgar Philipp Stoffel. A hybrid spatial model for representing indoor environments. In Web & Wireless Geographical Information Systems, International Symposium, W2gis, Hong Kong, China, December, pages 1–11, 2006.

161. Jan Oliver Wallgrün. Autonomous construction of hierarchical voronoi-based route graph representations. In International Conference on Spatial Cognition, pages 413–433. Springer, 2004.

162. Liping Yang and Michael Worboys. Generation of navigation graphs for indoor space. International Journal of Geographical Information Science, 29(10):1737–1756, 2015.

163. Liu Liu and Sisi Zlatanova. An approach for indoor path computation among obstacles that considers user dimension. ISPRS International Journal of Geo-Information, 4(4):2821–2841, 2015.

164. Jean-Claude Thill, Thi Hong Diep Dao, and Yuhong Zhou. Traveling in the three-dimensional city: applications in route planning, accessibility assessment, location analysis and beyond. Journal of Transport Geography, 19(3):405–421, 2011.

165. Abdullah Alattas, Sisi Zlatanova, Peter van Oosterom, and Ki-Joune Li. Improved and more complete conceptual model for the revision of IndoorGML. In Proceedings of the 10th International Conference on Geographic Information Science (GIScience 2018), volume 114, pages 21:1–21:12. Leibniz International Proceedings in Informatics, 2018.

166. buildingSMART International Ltd. Industry Foundation Classes Version 4.2 bSI Draft Standard IFC Bridge proposed extension. https://standards.buildingsmart.org/IFC/DEV/IFC4_2/FINAL/HTML/, 1996-2019. Accessed December 17, 2019.

167. Abdullah Kara, Volkan Cagdas, Ümit Isikdag, and Bülent Onur Turan. Towards harmonizing property measurement standards. Journal of Spatial Information Science, 2018(17):87–119, 2018.

168. Ya-Hong Lin, Yu-Shen Liu, Ge Gao, Xiao-Guang Han, Cheng-Yuan Lai, and Ming Gu. The ifc-based path planning for 3d indoor spaces. Advanced Engineering Informatics, 27(2):189–205, 2013.

169. Mikkel Boysen, Christian de Haas, Hua Lu, and Xike Xie. A journey from ifc files to indoor navigation. In International Symposium on Web and Wireless Geographical Information Systems, pages 148–165. Springer, 2014.

170. Shengjun Tang, Qing Zhu, Weixi Wang, and Yeting Zhang. Automatic topology derivation from ifc building model for indoor intelligent navigation. The International Archives of Photogrammetry, Remote Sensing and Spatial Information Sciences, 40(4):7, 2015.

171. Gerhard Gröger, Thomas H Kolbe, Claus Nagel, and Karl-Heinz Häfele. OGC City Geography Markup Language (citygml) encoding standard. `https://portal.opengeospatial.org/files/?artifact_id=47842`, 2012. Accessed June 10, 2020.

172. Tatjana Kutzner, Kanishk Chaturvedi, and Thomas H Kolbe. Citygml 3.0: New functions open up new applications. PFG–Journal of Photogrammetry, Remote Sensing and Geoinformation Science, pages 1–19, 2020.

173. Affan Idrees, Zahid Iqbal, and Maria Ishfaq. An efficient indoor navigation technique to find optimal route for blinds using qr codes. In 2015 IEEE 10th Conference on Industrial Electronics and Applications (ICIEA), pages 690–695. IEEE, 2015.

174. L' Ilkovicova, J Erdélyi, and A Kopacik. Positioning in indoor environment using qr codes. In INGEO 2014 – 6th International Conference on Engineering Surveying, Prague, Czech republic, April 3-4, 2014, pages 117–122, 2014.

175. José Antonio Puértolas Montañés, Adriana Mendoza Rodríguez, and Iván Sanz Prieto. Smart indoor positioning/location and navigation: A lightweight approach. International Journal of Artificial Intelligence and Interactive Multimedia, 2(2):43–50, 2013.

176. Sujith Suresh, PM Rubesh Anand, and D Sahaya Lenin. A novel method for indoor navigation using qr codes. International Journal of Applied Engineering Research, 10(77):2015, 2015.

177. G Venkatachalam, P Nivetha, M Keerthiga, and T Prema. QR code generation for mall shopping guide system with security. Asian Journal of Applied Science and Technology (AJAST), 1(4):37–39, 2017.

178. Ying Zhuang, Yuhao Kang, Lina Huang, and Zhixiang Fang. A geocoding framework for indoor navigation based on the qr code. In 2018 Ubiquitous Positioning, Indoor Navigation and Location-Based Services (UPINLBS), pages 1–4. IEEE, 2018.

179. Sung Hyun Jang. A qr code-based indoor navigation system using augmented reality. In GIScience–Seventh International Conference on Geographic Information Science, USA, 2012.

180. Josymol Joseph. Qr code based indoor navigation with voice response. International Journal of Science and Research, 3(11):923–926, 2014.

181. Wu You-bao and Xu Jian-min. The indoor precise location and navigation system based on two-dimensional code and A* algorithm. Electronic Design Engineering, 24(23):23–28, 2016.

182. David F Watson. Computing the n-dimensional delaunay tessellation with application to voronoi polytopes. The Computer Journal, 24(2):167–172, 1981.

183. Sisi Zlatanova, Jinjin Yan, Yijing Wang, Abdoulaye A. Diakité, Umit Isikdag, G Sithole, and Jack Barton. Spaces in spatial science and urban applications—state of the art review. ISPRS International Journal of Geo-Information, 9(1):58, 2020.

184. Alexis Richard C Claridades and Jiyeong Lee. Developing a data model of indoor points of interest to support location-based services. Journal of Sensors, 2020, 2020.

185. Qing Xiong, Qing Zhu, Sisi Zlatanova, Zhiqiang Du, Yeting Zhang, and L Zeng. Multi-level indoor path planning method. The International Archives of Photogrammetry, Remote Sensing and Spatial Information Sciences, 40(4):19, 2015.

186. Hosna Tashakkori, Abbas Rajabifard, and Mohsen Kalantari. A new 3d indoor/outdoor spatial model for indoor emergency response facilitation. Building and Environment, 89:170–182, 2015.

187. Stephan Winter and Yunhui Wu. Towards a conceptual model of talking to a route planner. In International Symposium on Web and Wireless Geographical Information Systems, pages 107–123. Springer, 2008.

188. Anna Labetski, Sarah van Gerwen, Guus Tamminga, Hugo Ledoux, and Jantien E. Stoter. A proposal for an improved transportation model in citygml. ISPRS - International Archives of the Photogrammetry, Remote Sensing and Spatial Information Sciences, pages 89–96, 2018.

189. Jinjin Yan, Sisi Zlatanova, and Abdoulaye A. Diakité. Two new pedestrian navigation path options based on semi-indoor space. ISPRS Annals of Photogrammetry, Remote Sensing and Spatial Information Sciences, VI-4/W1-2020:175–182, 2020.

190. Liu Liu, Sisi Zlatanova, Bofeng Li, Peter Van Oosterom, Hua Liu, and Jack Barton. Indoor routing on logical network using space semantics. ISPRS International Journal of Geo-Information, 8(3):126, 2019.

191. Jinjin Yan, Sisi Zlatanova, and Abdoulaye A. Diakité. Two new pedestrian navigation path options based on semi-indoor space. ISPRS Annals of Photogrammetry, Remote Sensing and Spatial Information Sciences, VI-4/W1-2020:175–182, 2020.

192. Reginald G Golledge. Path selection and route preference in human navigation: A progress report. In International Conference on Spatial Information Theory, pages 207–222. Springer, 1995.

193. Edsger W Dijkstra. A note on two problems in connexion with graphs. Numerische mathematik, 1(1):269–271, 1959.

194. Wei Zeng and Richard L Church. Finding shortest paths on real road networks: the case for a. International journal of geographical information science, 23(4):531–543, 2009.

195. Sudha Rani Kolavali and Shalabh Bhatnagar. Ant colony optimization algorithms for shortest path problems. In International Conference on Network Control and Optimization, pages 37–44. Springer, 2008.

196. Steven M LaValle et al. Rapidly-exploring random trees: A new tool for path planning. 1998.

197. Iram Noreen, Amna Khan, and Zulfiqar Habib. A comparison of RRT, RRT* and RRT*-smart path planning algorithms. International Journal of Computer Science and Network Security (IJCSNS), 16(10):20, 2016.

198. Durgesh Nandini and KR Seeja. A novel path planning algorithm for visually impaired people. Journal of King Saud University-Computer and Information Sciences, 31(3):385–391, 2019.

199. Matt Duckham and Lars Kulik. “simplest” paths: automated route selection for navigation. In International Conference on Spatial Information Theory, pages 169–185. Springer, 2003.

200. Zhiyong Wang and Sisi Zlatanova. Safe route determination for first responders in the presence of moving obstacles. IEEE Transactions on Intelligent Transportations Systems, pages 1–19, 2019.

201. Monir H Sharker, Hassan A Karimi, and Janice C Zgibor. Health-optimal routing in pedestrian navigation services. In Proceedings of the First ACM SIGSPATIAL International Workshop on Use of GIS in Public Health, pages 1–10. ACM, 2012.

202. Md Saniul Alam, Harikishan C. Perugu, and Aonghus McNabola. A comparison of route-choice navigation across air pollution exposure, co2 emission and traditional travel cost factors. Transportation Research Part D: Transport and Environment, 65:82–100, 2018.

203. Mandell Bellmore and George L Nemhauser. The traveling salesman problem: a survey. Operations Research, 16(3):538–558, 1968.

204. Eduardo L Pasiliao, Panos M Pardalos, and Leonidas S Pitsoulis. Branch and bound algorithms for the multidimensional assignment problem. Optimization Methods and Software, 20(1):127–143, 2005.

205. David R Morrison, Sheldon H Jacobson, Jason J Sauppe, and Edward C Sewell. Branch-and-bound algorithms: A survey of recent advances in searching, branching, and pruning. Discrete Optimization, 19:79–102, 2016.

206. Michael Jünger, Gerhard Reinelt, and Giovanni Rinaldi. The traveling salesman problem. Handbooks in Operations Research and Management Science, 7:225–330, 1995.

207. Keqi Zhang, Shu-Ching Chen, Dean Whitman, Mei-Ling Shyu, Jianhua Yan, and Chengcui Zhang. A progressive morphological filter for removing nonground measurements from airborne lidar data. IEEE Transactions on Geoscience and Remote Sensing, 41(4):872–882, 2003.

208. Hugo Ledoux and Martijn Meijers. Extruding building footprints to create topologically consistent 3d city models. In Urban and Regional Data Management, pages 51–60. CRC Press, 2009.

209. Filip Biljecki, Jantien Stoter, Hugo Ledoux, Sisi Zlatanova, and Arzu Çöltekin. Applications of 3d city models: State of the art review. ISPRS International Journal of Geo-Information, 4(4):2842–2889, 2015.

210. Aida E Afrooz, Russell Lowe, Simone Zarpelon Leao, and Chris Pettit. 3d and virtual reality for supporting redevelopment assessment. In Real Estate and GIS, pages 162–185. Routledge, 2018.

211. George Vosselman and Sander Dijkman. 3d building model reconstruction from point clouds and ground plans. International Archives of Photogrammetry Remote Sensing and Spatial Information Sciences, 34(3/W4):37–44, 2001.

212. Abdoulaye A. Diakité and Sisi Zlatanova. Automatic geo-referencing of bim in gis environments using building footprints. Computers, Environment and Urban Systems, 80, 2020.

213. Christiaan Lemmen, Peter Van Oosterom, and Rohan Bennett. The land administration domain model. Land Use Policy, 49:535–545, 2015.

214. Sisi Zlatanova, KJ Li, C Lemmen, and PJM van Oosterom. Indoor abstract spaces: linking IndoorGML and LADM. In Proceedings of the 5th International FIG Workshop on 3D Cadastres, 18-20 October 2016, Athens, Greece. Delft: FIG, 2016, pages 317–328, 2016.

215. Sisi Zlatanova, Peter Van Oosterom, J Lee, KJ Li, and CHJ Lemmen. LADM and IndoorGML for support of indoor space identification. ISPRS Annals of the Photogrammetry, Remote Sensing and Spatial Information Sciences, 4:257–263, 2016.

216. Yijing Wang, Yuning Cheng, Sisi Zlatanova, and Elisa Palazzo. Identification of physical and visual enclosure of landscape space units with the help of point clouds. Spatial Cognition & Computation, 20(3):257–279, 2020.

217. Bin Jiang and Xintao Liu. Automatic generation of the axial lines of urban environments to capture what we perceive. International Journal of Geographical Information Science, 24(4):545–558, 2010.

218. Norbert Haala and Claus Brenner. Generation of 3d city models from airborne laser scanning data. In 3rd EARSEL Workshop on LIDAR Remote Sensing on Land and Sea, Tallinn/Estonia, pages 105–112, 1997.

219. Vivek Verma, Rakesh Kumar, and Stephen Hsu. 3d building detection and modeling from aerial lidar data. In 2006 IEEE Computer Society Conference on Computer Vision and Pattern Recognition (CVPR'06), volume 2, pages 2213–2220. IEEE, 2006.

220. Christian Fruh and Avideh Zakhor. 3D model generation for cities using aerial photographs and ground level laser scans. In Proceedings of the 2001 IEEE Computer Society Conference on Computer Vision and Pattern Recognition. CVPR 2001, volume 2, pages II–II. IEEE, 2001.

221. Feifei Xie, Zongjian Lin, Dezhu Gui, and Hua Lin. Study on construction of 3d building based on uav images. The International Archives of the Photogrammetry, Remote Sensing and Spatial Information Sciences, 39:B1, 2012.

222. Shayan Nikoohemat, Abdoulaye A. Diakité, Sisi Zlatanova, and George Vosselman. Indoor 3d reconstruction from point clouds for optimal routing in complex buildings to support disaster management. Automation in Construction, 113:103109, 2020.

223. Christiaan Lemmen, Peter Van Oosterom, Rod Thompson, João Paulo Hespanha, and Harry Uitermark. The modelling of spatial units (parcels) in the land administration domain model (ladm). In Proceedings of the XXIV FIG International Congress 2010: Facing the Challenges - Building the Capacity, pages 1–21. International Federation of Surveyors (FIG), 2010.

224. Abdullah Alattas, Peter van Oosterom, Sisi Zlatanova, Abdoulaye A. Diakité, and Jinjin Yan. Developing a database for the LADM-IndoorGML model. In Proceedings of the 6th International FIG 3D Cadastre Workshop, 2-4 October 2018, Delft, The Netherlands, pages 261–278, 2018.

225. Abdullah Alattas, Peter van Oosterom, and Sisi Zlatanova. Deriving the Technical Model for the Indoor Navigation Prototype based on the Integration of IndoorGML and LADM Conceptual Model. In Proceedings of the 7th Land Administration Domain Model Workshop, Zagreb, page 24, 2018.

226. Abdullah Alattas, Peter van Oosterom, Sisi Zlatanova, Dick Hoeneveld, and Edward Verbree. Using the combined ladm-indoorgml model to support buiilding evacuation. ISPRS International Archives of the Photogrammetry, Remote Sensing and Spatial Information Sciences, XLII-4:11–23, 2018.

227. Abdullah Alattas, Sisi Zlatanova, Peter Van Oosterom, Efstathia Chatzinikolaou, Christiaan Lemmen, and Ki-Joune Li. Supporting indoor navigation using access rights to spaces based on combined use of indoorgml and ladm models. ISPRS International Journal of Geo-Information, 6(12):384, 2017.

228. Ann Vanclooster, Nina Vanhaeren, Pepijn Viaene, Kristien Ooms, Laure De Cock, Veerle Fack, Nico Van De Weghe, and Philippe De Maeyer. Turn calculations for the indoor application of the fewest turns path algorithm. International Journal of Geographical Information Science, 33(11):2284–2304, 2019.

Index

2D indoor navigation models, 46–48
3D navigation models, 48, 49
3D space models
reconstructing approach, 87
reconstruction of building shells, 100–111
sI-space reconstruction, 87–94
sO-spaces and O-spaces, 94–100

A
Absolute space, 22
AbstractSpaceBoundary, 68
reconstruction, 50, 61, 67, 87, 118–120, 144
AbstractUnoccupiedSpace, 68
Agents, 6–8
Air navigation, 1
Atriums, 56

B
Balconies, 30
Barriers, 24
BIM, *see* Building Information Model
BottomClosure, 13
Boundaries, 15, 16
Boundary representation (BRep), 32
BoundedBy property, 51, 52
BRep, *see* Boundary representation
Building elements
bottom, 12
closure, 12–15
side, 11, 12
top, 11
Building footprint, 8, 9, 100–104, 121–123
Building height, 9, 10
Building Information Model (BIM), 33, 51, 116, 118
Building Installation class, 51
Building shells reconstruction
3D building model
and DTM, 108–110
and point clouds, 110, 111
algorithms, 105–107
DTM, 101
footprints and point clouds, 107, 108
generate top and bottom, 102
illustration, 102–105
point clouds, 111–113
rebuild terrain, 102
set height and create sides, 101, 102
TIC computation process, 101

C
CAD, *see* Computer-aided Design
CDT, *see* Constrained Delaunay triangulation
Ceiling Surface, 52
Cell Space reconstruction, 50, 68, 87, 119, 120, 144, 146
City Geography Markup Language (CityGML), 51–53
Closure Surface class, 52, 69
Computer-aided Design (CAD), 33
Constrained Delaunay triangulation (CDT), 10, 55, 102
Constrained triangulation representing the terrain, 102
Contemporary public buildings, 141
Coordinate reference system (CS), 100, 115
Corridors, 30, 47, 56, 83, 130
CS, *see* Coordinate reference system

D
Data preparation, 115
Database management system (DBMS), 70, 120
DBMS, *see* Database management system
Digital Elevation Model (DEM), 10
Digital Ground Model (DGM), 10

Digital Height Model (DHM), 10
Digital Terrain Elevation Models (Stems), 10
Digital Terrain Model (DTM), 10, 100, 109
Dijkstra algorithm, 71
DTM, *see* Digital Terrain Model
Dummy nodes, 58–60

E

EM, *see* Estate Management
Entrance node, 4
Environments, 142, 143
Envision Scenarios Planner, 9
Estate Management (EM), 115
Extended navigation network, 58, 59

F

Faces, 32
Feature Manipulation Engine (FME), 105, 115
FME, *see* Feature Manipulation Engine
FSS framework, *see* Flexible Space Subdivision framework
Flexible Space Subdivision framework (FSS framework), 147
Floor Surface, 52
Footprints, 8, 9

G

Generators, 55
Generic spaces
  bridges between two buildings, 44
  built scenes, 42
  descriptive definition, 35, 36
  dimensions and closure, 43
  flowchart, 39
  weak characteristics, 43
Geographic Information Systems (GIS), 33
Geometric representations
  3D volumes, 33
  building micro-climate, 34
  IndoorGML, 34
  landscape, 34
  outdoor related research fields, 35
GIS, *see* Geographic Information Systems
Global Navigation Satellite System (GNSS), 4
GNSS, *see* Global Navigation Satellite System
Gradient, 15
GroundSurface, 51

I

IAI, *see* International Alliance for Interoperability
Indoor (I-space), 21–23, 26, 27, 36
Indoor navigation path options, 72
Indoor POI, 56
Indoor Positioning System (IPS), 4
Indoor scene classification, 55–57
Indoor travelling salesman problem (ITSP), 71, 80, 81
  concepts and modeling, 82
  data preprocessing, 130, 131
  illustration, 83–85
  navigation network derivation, 131–135
  planning, 135–139
  procedures, 82, 83
  undirected graph, 83, 85
IndoorGML navigation model, 50, 51
Industry Foundation Classes (IFC), 51
Inertial Navigation System (INS), 4
InteriorWallSurface, 52
International Alliance for Interoperability (IAI), 51
International standards
  CityGML, 51–53
  IFC, 51
  IndoorGML, 49, 50
IPS, *see* Indoor Positioning System
ITSP, *see* Indoor travelling salesman problem

L

Land Administration Data Model (LADM), 142, 150

Land navigation, 1
Landscape spaces, 34
Living environments, 25, 26
Location-based services (LBS), 70

M
Marine navigation, 1
Media Axis Transform (MAT), 47
Most-Top-Covered path (MTC-path), 71
MTC-path, 74, 75

N
National Elevation Data Framework (NEDF), 116
NavigableBoundary, 50
NavigableSpace, 50
Navigation
- application, 2
- components, 3
- definition, 1, 2
- seamless, 2, 3 (*see also* Seamless navigation)

Navigation network derivation
- 2D, 46–48
- 3D, 48, 49
- ITSP-path, 131–135
- QR codes, 53–55, 60
- space selection, 122–124
- space subdivision, 57, 58, 61

Network models, 45, 46
Node-Relation Graph (NRG), 46
Non-cuboid spaces, 18
NSI-path, 75, 76

O
Obstacles, 57
Open Geospatial Consortium (OGC), 49
OpenStreetMap (OSM), 9
Outdoor (O-space), 36, 44, 142
Outdoor navigation, 2
Outdoor space, 21–23, 26, 27
OuterCeilingSurface, 51
OuterFloorSurface, 51

P
Path options
- comparing paths, 80
- illustration, 77–80
- indoor, 72
- ITSP-path, 80, 81 (*see also* Indoor travelling salesman problem (ITSP))
- modified weights, 78
- MTC-path, 74, 75
- navigation graph, 78
- for navigation model
  - covered/uncovered distance, 73
  - distance between two connected spaces, 72, 73
  - modified weights, 73
  - original weights, 73
  - uncovered ratio, 73
- for navigation path
  - covered/uncovered length of a path, 74
  - path length, 74
  - top-coverage-ratio of a path, 74
  - total weight of a path, 74
- navigation paths, 79
- NSI-path, 75, 76
- selection strategy, 76, 77

Path planning
- comparing navigation systems, 129, 130
- MTC-path, 126–129
- NSI-path, 126–129
- seamless navigation, 124–126

Path to the Nearest Semi-indoor (NSI-path), 71
Path-finding, *see* Navigation
Pedestrian navigation, 1, 4
Physical boundary, 15
Poincaré duality theory, 46
Point of Interest (POI), 54, 56
Projecting the whole footprint Polyline on the Terrain (PPT), 104
Projecting Vertices of footprints on the Terrain (PVT), 101

Q
QR, *see* Quick Response code
QR code-based indoor navigation, 53–55
Quantitative definitions, 37, 38
Quantum GIS (QGIS), 115, 130, 131
Quick Response code (QR), 53–55

R
Radio Frequency Identification (RFID), 4
Radio signal strength information (RSSI), 53
Rapidly-exploring random tree (RRT), 71
Reconstruction of building shells, 100–111
Relational Database Management Systems (RDBMS), 66
RoofSurface, 51
Room-like space, 56

S
Seamless navigation, 2
  anchor node, 5
  indoor, 4
  outdoor, 5
  resolutions, 5
Seamless path planning, 141
Semi-bounded environments
  with upper boundaries, 26
  without upper boundaries, 27
Semi-bounded spaces, 16, 27, 28
Semi-enclosed spaces, 27
Semi-indoor (sI-space), 5, 29, 36
Semi-outdoor (sO-spaces), 5, 29, 36
SI-space reconstruction, 87
  algorithms, 91–94
  determination of top and bottom, 88–90
  identified building components, 88
  illustration, 90, 91
  projections, 89
  space generation, 88–90
  trimming spaces, 90
SideClosure, 13
Sites, 55
SO-spaces and O-spaces
  3D space models, 94–100
  algorithms, 98–100
  classifying, 95, 96
  extracting footprints, 94, 95
  illustration, 96–98
  reconstruct 3D spaces, 96
Space generation, 88–90
Space height, 18
Space navigation, 1
Space radius, 18
Space subdivision, 57, 58, 61, 147
Spaces
  boundaries, 15, 16
  categories, 30
  classification and reconstruction, 118–122
  classification based on building, 30
  computation of vertices, 123
  definitions, 21–23
  descriptive definitions, 36
  developments, 144, 145
  geometric representations
    3D volumes, 33
    building micro-climate, 34
    IndoorGML, 34
    landscape, 34
    outdoor related research fields, 35
  gradient, 15
  physical boundaries, 120
  quantitative definitions, 37, 38, 41
  radius/height, 16–18
  reception of satellites signal, 31
  representation
    BRep, 32
    voxels, 32, 33
  research
    data sources, 148, 149
    extensions, 146
    investigate space accessibility, 149, 150
    multi-purpose uses, 148

navigation performance, 147, 148
new navigation path options, 150
obstacles, 147
space subdivision, 147
threshold values, 37–40
type of signal received by the users, 31
types, 28
urban applications, 24
visual aid, 31
Spaces representation, 143
SType distinguishes spaces, 65
Surrounding planes, 146

T

Tacking path, 2
Terrain, 10
Terrain intersection curve (TIC), 6, 10
TopClosure, 12
Traffic spaces, 52, 53
Transformer, 105
Traveling salesman problem (TSP), 71, 81
Triangulated Irregular Network (TIN), 10
Trimming spaces, 90

U

Ultra Wide Band (UWB), 61
Unified 3D space-based navigation model (U3DSNM), 63
accessibility of spaces, 69
attributes, 65
CityGML 3.0, 66–69
closures, 65
conceptual model, 64–66
IndoorGML 1.0, 66–69
PhysicalBoundary table, 67
python classes, 66
requirements, 63, 64
sI-space table, 66
technical model, 66
UML diagram, 65
use cases, 64
Unified Modeling Language (UML), 65
Unified navigation model, 4, 63, 143, 144
University of New South Wales (UNSW), 115, 116

V

Vehicle navigation, 2
Virtual boundary, 15
Voronoi diagram, 48, 55
Voronoi tessellation, 59
Voxels, 32, 33

W

WallSurface, 52
Way-finding, 1